Solutions Manual for
Guide to Energy Management,
Eighth Edition
International Version

T0321031

Solutions Manual for
Guide to Energy Management,
Eighth Edition
International Version

Klaus-Dieter E. Pawlik

THE FAIRMONT PRESS, INC.

CRC Press
Taylor & Francis Grou

Solutions Manual for Guide to Energy Management, Eighth Edition, International Version / By Klaus-Dieter E. Pawlik.

Published by The Fairmont Press, Inc.
700 Indian Trail
Lilburn, GA 30047
tel: 770-925-9388; fax: 770-381-9865
http://www.fairmontpress.com

Distributed by Taylor & Francis Group LLC
6000 Broken Sound Parkway NW, Suite 300
Boca Raton, FL 33487, USA
E-mail: orders@crcpress.com

Distributed by Taylor & Francis Group LLC
23-25 Blades Court
Deodar Road
London SW15 2NU, UK
E-mail: uk.tandf@thomsonpublishingservices.co.uk

Printed in the United States of America
10 9 8 7 6 5 4 3 2 1

ISBN-0-88173-774-7 (The Fairmont Press, Inc., Print Version)
ISBN-0-88173-775-9 (The Fairmont Press, Inc., Electronic Version)
ISBN-978-1-4987-7989-0 (Taylor & Francis Group LLC)

Table of Contents

Chapter 1

Introduction to Energy Management

Problem: For your university or organization, list some energy management projects that might be good "first ones," or early
selections.

Solution: Early projects should have a rapid payback, a high probability of success, and few negative consequences (increas-
ing/decreasing the air-conditioning/heat, or reducing
lighting levels).

Examples:

Switching to a more efficient light source (especially
in conditioned areas where one not only saves with
the reduced power consumption of the lamps but also
from reduced refrigeration or air-conditioning load).

Repairing steam leaks. Small steam leaks become large
leaks over time.

Insulating hot fluid pipes and tanks.

Install high efficiency motors.

And many more

Problem: Again for your university or organization, assume you are starting a program and are defining goals. What are some potential first-year goals?

Solution: Goals should be tough but achievable, measurable, and specific.

Examples:
Total energy per unit of production will drop by 10 percent for the first six months and an additional 5 percent the second half of the year.

Within 2 years all energy consumers of 5 million kilojoules per hour (kJ/h) or larger will be separately metered for monitoring purposes.

Each plant in the division will have an active energy management program by the end of the first year.

All plants will have contingency plans for gas curtailments of varying duration by the end of the first year.

All boilers of 25,000 kg/h or larger will be examined for waste heat recovery potential the first year.

Problem: If you were a member of the upper level management in charge of implementing an energy management program at your university or organization, what actions would you take to reward participating individuals and to reinforce commitment to energy management?

Solution: The following actions should be taken to reward individuals and reinforce commitment to energy management:

Develop goals and a way of tracking their progress.

Develop an energy accounting system with a performance measure such as kJ/m^2 or $kJ/unit$.

Assign energy costs to a cost center, profit center, an investment center or some other department that has an individual responsibility for cost or profit.

Reward (with a monetary bonus) all employees who control cost or profit relative to the level of cost or profit. At the risk of being repetitive, note that the level of cost or profit should include energy costs.

Problem: Perform the following energy conversions and calculations:

a) A spherical balloon with a diameter of three meters is filled with natural gas. How much energy is contained in that quantity of natural gas?

b) How many Joules are in 550 cubic metres of natural gas? How many GJ in 2,000 litres of #2 fuel oil?

c) An oil tanker is carrying 3,000 litres of #2 fuel oil. If each litre of fuel oil will generate 3.3 kWh of electric energy in a power plant, how many kWh can be generated from the oil in the tanker?

d) How much hard coal is required at a power plant with a heat rate of 10 MJ/kWh to run a 6 kW electric resistance heater constantly for 1 week (168 hours)? (One tonne of hard coal conttains 25 GJ of heat.)

e) A large city has a population which is served by a single electric utility which burns hard coal to generate electrical energy. If there are 500,000 utility customers using an average of 12,000 kWh per year, how many tonnes of coal must be burned in the power plants if the heat rate is 10.5 MJ/kWh? (One tonne of hard coal contains 25 GJ of heat.)

f) Consider an electric heater with a 4,500 watt heating element. Assuming that the water heater is 98% efficient, how long will it take to heat 200 litres of water from 20 degrees C to 60 degrees C?

Solution:

a) V = 4/3 (PI) r^3

= 4/3 × 3.14 × 1.5^3

= 14.13 m^3

E = V × 38.14 MJ/m^3 of natural gas

= 14.13 m^3 × 38.14 MJ/m^3

= *539 MJ*

b) E = 550 m^3 × 38.14 MJ/m^3 of natural gas × 1,000,000 J/MJ

= *2.10E+10 J*

E = 2,000 L × 39 MJ/L of #2 fuel oil × GJ/1,000 MJ

= *78 GJ*

c) E = 3,000 L × 3.3 kWh/L

= *9,000 kWh*

d) V = 10,000 MJ/kWh × 6 kW × (168 h/25 GJ/tonne hard coal) × (GJ/1,000 MJ)

= *0.40 tonnes of coal*

e) V = 500,000 cus. × 12,000 kWh/cus. × 10.5 MJ/kWh × (1 tonne/25 GJ) × (GJ/1,000 MJ)

= *2,520,000 tonnes of coal*

f) ΔQ = cmΔT

= specific heat constant × mass × change in temperature

= (4.186 kJ/kg/C) × 200 L × (0.001 m^3/L) × (998 kg/m^3) × (60 C – 20 C)

= 33,421 kJ

p = 4,500 W × 1 kW/1,000 W × 1 kJ/s/kW

= 4.5 kJ/s

t = ΔQ/p/efficiency

= 33,421 kJ/(4.5 kJ/s) × (1h/3,600 s)/0.98

= *2.11 h*

Problem: A person takes a shower for ten minutes. The water flow rate is 12 litres per minute, the temperature of the shower water is 45 degrees C. Assuming that cold water is at 16 degrees C, and that hot water from a 70% efficient gas water heater is at 60 degrees C, how many cubic metres of natural gas does it take to provide the hot water for the shower?

Solution: ΔQ = cmΔT eff
      = specific heat constant × mass × change in temperature
      = (4.186 kJ/kg/C) × 10 min × 12 L/min × (0.001 m^3/L) × (998 kg/m^3) × (45 C – 16 C)/0.7
      = 20,769 kJ

   V = (20,769 kJ/38.14 MJ/m^3) × MJ/1,000 kJ
     = *0.54 m^3 of natural gas*

Problem: An office building uses 1 million kWh of electric energy and 12,000 litres of #2 fuel oil per year. The building has 4,000 square metres of conditioned space. Determine the energy use index (EUI) and compare it to the average EUI of an office building.

Solution: E(elect.) = 1,000,000 kWh/yr. × kJ/s/kW × 3,600 s/h
                              = 3,600,000,000 kJ/yr.

                E(#2 fuel) = 12,000 L/yr. × 39 MJ/L × 1,000 kJ/MJ
                              = 468,000,000 kJ/yr.

                         E = 4,068,000,000 kJ/yr.

                     EUI = 4,068,000,000 kJ/yr./4,000 m²

                          = *1,017,000 kJ/m²/yr. which is
                             less than the average office building*

Problem: The office building in Problem 1.6 pays €65,000 a year for electric energy and €9,900 a year for fuel oil. Determine the energy cost index (ECI) for the building and compare it to the ECI for an average building.

Solution: ECI $= (€65,000 + €9,900)/4,000 \ m^2$
               $= €18.73/m^2/yr.$
               which is greater than the average building

Problem:	As a new energy manager, you have been asked to pre-dict the energy consumption for electricity for next month (February). Assuming consumption is dependent on units produced, that 1,000 units will be produced in February, and that the following data are representative, determine your estimate for February.		

	Month	Units produced	Consumption (kWh)	Average (kWh/unit)	
Given:					
	January	600	600	1.00	
	February	1,500	1,200	0.80	
	March	1,000	800	0.80	
	April	800	1,000	1.25	
	May	2,000	1,100	0.55	
	June	100	700	7.00	Vacation month
	July	1,300	1,000	0.77	
	August	1,700	1,100	0.65	
	September	300	800	2.67	
	October	1,400	900	0.64	
	November	1,100	900	0.82	
	December	200	650	3.25	1-week shutdown
	January	1,900	1,200	0.63	

Solution: First, since June and December have special circumstances, we ignore these months. We then run a regression to find the slope and intercept of the process model. We assume that with the exception of the vacation and the shutdown that nothing other then the number of units produced affects the energy used. Another method of solving this problem may assume that the weather and temperature changes also affect the energy use.

Month	Units produced	Consumption (kWh)	Average (kWh/unit)
January	600	600	1.00
February	1,500	1,200	0.80
March	1,000	800	0.80
April	800	1,000	1.25
May	2,000	1,100	0.55
July	1,300	1,000	0.77
August	1,700	1,100	0.65
September	300	800	2.67
October	1,400	900	0.64
November	1,100	900	0.82
January	1,900	1,200	0.63

From the ANOVA table, we see that if this process is modeled linearly the equation describing this is as follows:

$$kWh\ (1{,}000\ units) = 623 + 0.28 \times kWh/unit\ produced$$
$$= \textbf{\textit{899 kWh}}$$

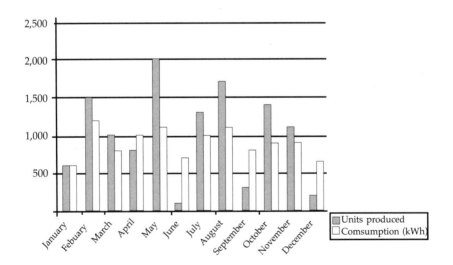

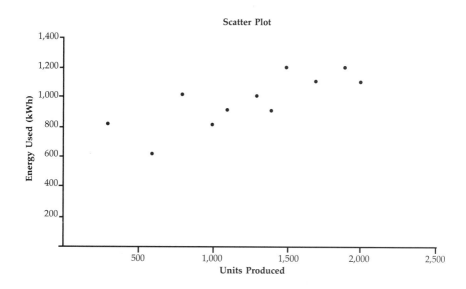

SUMMARY OUTPUT

Regression Statistics

Multiple R	0.795822426
R Square	0.633333333
Adjusted R Square	0.592592593
Standard Effort	118.6342028
Observations	11

ANOVA

	df	SS	MS	F	Significance F
Regression	1	218787.9788	218787.9	15.54545	0.00339167
Residual	9	126666.6667	14074.07		
Total	10	345454.5455			

	Coefficients	Standard Error	t Stat	P-value	Lower 95%	Upper 95%	Lower 95. 0%	Upper 95.0%
Intercept	623.1884058	93.46296795	6.667759	9.19E-05	411.7603222	834.616489	411.760322	834.6164893
X Variable 1	0,275362319	0.06993977	3.942772	0.003392	0,117373664	0.43335097	0.11737366	0.433350974

Problem: For the same data as given in Problem 1.8, what is the fixed energy consumption (at zero production, how much energy is consumed and for what is that energy used)?

Solution: By looking at the regression run for problem 1.8 (see ANOVA table), we can see the intercept for the process in question. This intercept is probably the best estimate of the fixed energy consumption:

623 kWh.

This energy is probably used for space conditioning and security lights.

Problem: Determine the cost of fuel switching, assuming there were 1,000 cooling degree days (CDD) and 1,000 units produced in each year.

Given: At the Gator Products Company, fuel switching caused an increase in electric consumption as follows:

	Expected energy consumption	Actual energy consumption after switching fuel
Electric/CDD	75 GJ	80 GJ
Electric/units of production	100 GJ	115 GJ

The base year cost of electricity is €30 per GJ, while this year's cost is €35 per million GJ

Solution: Cost variance = €35/GJ – €30/GJ
                             = €5/GJ

Increase cost due to cost variance
    = Cost variance × Total Actual Energy Use
    = (€5/GJ) × ((80 GJ/CDD) × (1,000 CDDs) +
    (115 GJ/unit) × (1,000 units))
    = €975,000

CDD electric variance
    = 1,000 CDD × (80 – 75) GJ/CDD
    = 5,000 GJ

Units electric variance
    = 1,000 units × (115 – 100) GJ/unit
    = 15,000 GJ

Increase in energy use
    = CDD electric variance + Units electric variance
    = 5,000 GJ + 15,000 GJ
    = 20,000 GJ

Increase cost due to increased energy use
= Increase in energy use × Base cost of electricity
= 20,000 GJ × €30/GJ
= €600,000

Total cost of fuel switching
= Increase cost due to increased energy use
+ Increased cost due to cost variance
= €600,000 + €975,000
= **€1,575,000**

Chapter 2

The Energy Audit Process: An Overview

Problem: Compute the number of heating degree days (HDD) associated with the following weather data.

	Time Period	Tempera-ture (degrees C)	Number of hours	18C -Tem-perature (degrees C)	Hours × dT
Given:	Midnight - 4:00 AM	−7	4	25	100
	4:00 AM - 7:00 AM	−10	3	28	84
	7:00 AM - 10:00 AM	−8	3	26	78
	10:00 AM - Noon	−6	2	24	48
	Noon - 5:00 PM	−1	5	19	95
	5:00 PM - 8:00 PM	−4	3	22	66
	8:00 PM - Midnight	−7	4	25	100
					571

Solution: From the added columns in the given table, we see that the number of hours times the temperature difference from 18 degrees C is 571 C-hours. Therefore, the number of HDD can be calculated as follows:

$$\text{HDD} = 571 \text{ C-hours}/24 \text{ h/day}$$
$$= \mathbf{23.79} \text{ degree-days}$$

Problem: Select a specific type of manufacturing plant and describe the kinds of equipment that would likely be found in such a plant.

List the audit data that would need to be collected for each piece of equipment.

What particular safety aspects should be considered when touring the plant?

Would any special safety equipment or protection be required?

Solution: The following equipment could be found in a wide variety of manufacturing facilities:

Equipment	Audit data
Heaters	Power rating
	Use characteristics (annual use, used in conjunction with what other equipment, how is the equipment used?)
Boilers	Power rating
	Use characteristics
	Fuel used
	Air-to-fuel ratio
	Percent excess air
Air-conditioners	Power rating
Chillers	Efficiency
Refrigeration	Cooling capacity
	Use characteristics
Motors	Power rating
	Efficiency
	Use characteristics
Lighting	Power rating
	Use characteristics
Air-compressors	Power rating
	Use characteristics
	Efficiency
	Various air pressures
	An assessment of leaks

Specific process equipment for example for a metal furniture plant one may find some sort of electric arc welders for which one would collect its power rating and use characteristics.

The following include a basic list of some of the safety precautions that may be required and any safety equipment needed:

Safety precaution   *Safety equipment*

> As a general rule of thumb the auditor should never touch any-thing—just collect data. If a measurement needs to be taken or equipment manipulated, ask the operator.

Beware of rotating machinery
Beware of hot machinery/pipes   *Asbestos gloves*
Beware of live circuits   *Electrical gloves*

> Have a trained electrician take any electrical measurements

Avoid working on live circuits, if possible.

Securely lock and tag circuits and switches in the off/open position before working on a piece of equipment.

Always keep one hand in your pocket while making measurements on live circuits to help prevent accidental electrical shocks.

When necessary, wear a full face respirator mask with adequate filtration particle size.

Use activated carbon cartridges in the mask when working around low concentrations of noxious gases. Change cartridges on a regular basis.

Use a self-contained breathing apparatus for work in toxic environments.

Use foam insert plugs while working around loud machinery to reduce sound levels by nearly 30 decibels (in louder environments hearing protection rated at higher noise levels may be required)

> Always ask the facility contact about special safety precautions or equipment needed. Additional information can be found in OSHA literature.

For our metal furniture plant:
Avoid looking directly   *Tinted safety goggles*
      at the arc of the welders

Problem: Section 2.1.2 of the *Guide to Energy Management* provided a list of energy audit equipment that should be used. However, this list only specified the major items that might be needed. In addition, there are a number of smaller items such as hand tools that should also be carried. Make a list of these other items, and give an example of the need for each item.

How can these smaller items be conveniently carried to the audit?

Will any of these items require periodic maintenance or repair?

If so, how would you recommend that an audit team keep track of the need for this attention to the operating condition of the audit equipment?

Solution: Smaller useful audit equipment may include:
A flashlight
Extra batteries
A hand-held tachometer
A clamp-on ammeter
Recording devices

These smaller items can be conveniently carried in a tool box.

As with most equipment, these items will require periodic maintenance. For example, the flashlight batteries and light bulbs will have to be changed.

For these smaller items, one could probably just include the periodic maintenance as part of a pre-audit checklist. For items that require more than just cursory maintenance, one could include the item in their periodic maintenance system.

Problem: Section 2.2 of the *Guide to Energy Management* discussed the point of making an inspection visit to a facility at several different times to get information on when certain pieces of equipment need to be turned on and when they are unneeded. Using your school classroom or office building as a specific example, list some of the unnecessary uses of lights, air conditioners, and other pieces of equipment. How would you recommend that some of these uses that are not necessary be avoided? Should a person be given the responsibility of checking for this unneeded use? What kind of automated equipment could be used to eliminate or reduce this unneeded use?

Solution: Typically, one could visit a university at night and observe that the lights of classrooms are on even at midnight when no one is using the area. One idea would be to make the security force responsible for turning off non-security lights when they make their security tours at night. A better idea may be to install occupancy sensors so that the lights are on only when the area is in use. An additional benefit of occupancy sensors could be security; many thieves or vandals would be startled when lights come on.

Problem: An outlying building has a 25 kW company-owned transformer that is connected all the time. A call to a local electrical contractor indicates that the core losses from comparable transformers are approximately 3% of rated capacity. Assume that the electrical costs are ten cents per kWh and €10/kW/month of peak demand, that the average building use is ten hours/month, and that the average month has 720 hours. Estimate the annual cost savings from installing a switch that would energize the transformer only when the building was being used.

Given:

Transformer power use	25	kW
Core losses	3%	
Electrical energy cost	€0.10	/kWh
Demand charge	€10.00	/kW/month
Building utilization	10	hrs/mo
Hours in a month	720	hrs/mo
Months in a year	12	mo/yr

Solution: The energy savings (ES) from installing a switch that would energize the transformer only when the building was being used can be calculated as follows:

$$\text{ES} = \text{(Percentage of core losses)(Transformer power use)(Hours in a month} - \text{Building utilization)(Months in a year)}$$
$$= 3\% \times 25 \text{ kW} \times (720 - 10) \text{ hrs/mo} \times 12 \text{ mo/yr}$$
$$= 6{,}390 \text{ kWh/yr}$$

Since we do not expect the monthly peak demand to be reduced by installing this switch, the only savings will come from energy savings. Therefore, annual savings (AS) can be calculated as follows:

$$\text{AS} = \text{ES} \times \text{Electrical energy cost}$$
$$= 6{,}390 \text{ kWh/yr} \times €0.10/\text{kWh}$$
$$= €639/yr$$

Chapter 3

Understanding Energy Bill

Problem: By periodically turning off a fan, what is the total euro savings per year to the company?

Given: In working with Ajax Manufacturing Company, you find six large exhaust fans are running constantly to exhaust general plant air (not localized heavy pollution). They are each powered by 25-kW electric motors with loads of 27 kW each. You find they can be turned off periodically with no adverse effects. You place them on a central timer so that each one is turned off for 10 minutes each hour. At any time, one of the fans is off, and the other five are running. The fans operate 10 h/day, 250 days/year. Assume the company is on the rate schedule given in Figure 3-10. Neglect any ratchet clauses. The company is on service level 3 (distribution service). (There may be significant HVAC savings since conditioned air is being exhausted, but ignore that for now.)

Solution: <u>Demand charge</u>

On-peak €12.22/kW/mo	June-October	5 months/year
Off-peak €4.45/kW/mo	November-May	7 months/year

<u>Energy charge</u>
For first two million kWh €0.03431/kWh
All kWh over two million €0.03010/kWh

<u>Assumptions (and possible explanations)</u>
Assume the company uses well over two million kWh per month.
The fuel cost adjustment is zero, since the utility's fuel cost is at the base rate.
There is no sales tax since the energy can be assumed to be used for production.

The power factor is greater than 0.8

No franchise fees since the company is outside any municipality

The demand savings (DS) can be calculated as follows:

$$DS = [(DC \text{ on peak}) \times (N \text{ on peak}) + (DC \text{ off peak}) \times (N \text{ off peak})] \times DR$$

where,

DC = Demand charge for specified period

N = Number of months in a specified period

DR = Demand reduction, 27 kW since a motor using this amount is always turned off with the new policy

Therefore,

$$DS = [(€12.22/\text{kW}/\text{mo}) \times (5 \text{ mo}/\text{yr}) + (€4.45/\text{kW}/\text{mo}) \times (7 \text{ mo}/\text{yr})] \times 27 \text{ kW} = €2{,}490.75/\text{yr}$$

The energy savings (ES) can be calculated as follows:

$$ES = (EC > 2 \text{ million}) \times (10 \text{ h}/\text{day}) \times (250 \text{ day}/\text{yr}) \times DR$$

where

EC = Marginal energy charge

Therefore,

$$ES = (€0.03010/\text{kWh}) \times (10 \text{ h}/\text{day}) \times (250 \text{ day}/\text{yr}) \times 27 \text{ kW}$$
$$= €2{,}031.75/\text{yr}$$

Finally, the total annual savings (TS) can be calculated as follows:

$$TS = DS + ES$$
$$= €4{,}522.50/yr$$

Additional Considerations

How much would these timers cost?

How much would it cost to install these timers? Or an alternate control system?

Does cycling these fans on and off cause the life of the fan motors to decrease?

What would the simple payback period be?

Net present value?

Internal rate of return?

Problem: What is the euro savings for reducing demand by 100 kW in the off-peak season?

             If the demand reduction of 100 kW occurred in the peak season, what would be the euro savings (that is, the demand in June through October would be reduced by 100 kW)?

Given: A large manufacturing company is on the rate schedule shown in Figure 3-10 (service level 5, secondary service). Their peak demand history for last year is shown below. Assume they are on the 65% ratchet clause specified in Figure 3-10. Assume the high month was July of the previous year at 1,150 kW.

Month	Demand (kW)	Month	Demand
Jan	495	*Jul*	*1100*
Feb	550	*Aug*	*1000*
Mar	580	*Sep*	*900*
Apr	600	*Oct*	*600*
May	610	Nov	500
Jun	*900*	Dec	515

Note italics indicates on-peak season

Solution: Demand charge

On-peak €13.27/kW/mo	June-October	5 months/year
Off-peak €4.82/kW/mo	November-May	7 months/year

Ratchet clause
Dpeak = max (actual demand corrected for pf, 65% of the highest on-peak season demand corrected for pf)

Assumptions (and possible explanations)
Assume the company uses well over two million kWh per month.
The fuel cost adjustment is zero, since the utility's fuel cost is at the base rate.

There is no sales tax since the energy can be assumed to be used for production.

The power factor is greater than 0.8.

No franchise fees since the company is outside any municipality.

Estimated next year with a 100 kW decrease in the off-peak season

Month	Demand (kW)	Ratchet	Euro savings
Jan	395	747.5	0
Feb	450	747.5	0
Mar	480	747.5	0
Apr	500	747.5	0
May	510	747.5	0
Jun	900	747.5	0
Jul	1100	715	0
Aug	1000	715	0
Sep	900	715	0
Oct	600	715	0
Nov	400	715	0
Dec	415	715	0
			0

Therefore, you would not save any money by reducing the peak demand in the off-season. This non-savings is due to the ratchet and the degree of unevenness of demand.

Estimated next year with a 100 kW decrease in the on-peak season

Month	Demand (kW)	Ratchet	Euro savings
Jan	495	747.5	0
Feb	550	747.5	0
Mar	580	747.5	0
Apr	600	747.5	0
May	610	747.5	0
Jun	800	747.5	€1,327
Jul	1000	650	€1,327
Aug	900	650	€1,327
Sep	800	650	€1,327
Oct	500	650	€863
Nov	500	650	€313
Dec	515	650	€313

The first year they would save: €6,797

Every year after that they would save the following:

$$\text{Savings} = 65\% \times 100 \text{ kW} \times 7 \text{ mo/yr} \times €4.82/\text{kW/mo}$$
$$+ 100 \text{ kW} \times 5 \text{ mo/yr} \times €13.27/\text{kW/mo}$$
$$= €8,828/yr$$

Problem: Use the data found in Problem 3.2. How many months would be ratcheted, and how much would the ratchet cost the company above the normal billing?

Solution: Assuming that the 100 kW reduction is not made

Month	Demand (kW)	Ratchet	Ratchet Cost
Jan	495	747.5	€1,217.05
Feb	550	747.5	€951.95
Mar	580	747.5	€807.35
Apr	600	747.5	€710.95
May	610	747.5	€662.75
Jun	900	747.5	€—
Jul	1100	715	€—
Aug	1000	715	€—
Sep	900	715	€—
Oct	600	715	€1,526.05
Nov	500	715	€1,036.30
Dec	515	715	€964.00

8 months would be ratcheted at a cost of €7,876.40

Problem: Calculate the savings for correcting to 80% power factor? How much capacitance (in kVARs) would be necessary to obtain this correction?

Given: In working with a company, you find they have averaged 65% power factor over the past year. They are on the rate schedule shown in Figure 3-10 and have averaged 1,000 kW each month. Neglect any ratchet clause and assume their demand and power factor are constant each month. Assume they are on transmission service (level 1).

Solution: <u>Demand Charge</u>

On-peak €10.59/kW/mo June-October   5 months/year
Off-peak €3.84/kW/mo   November -May 7 months/year

Billed Demand = Actual Demand × (base pf/actual pf)
               = 1000 kW × 0.8/0.65
               = 1231 kW

pf correction savings = 231 kW × (5 mo/yr × €10.59/kW/mo
                         + 7 mo/yr × €3.82/kW/mo)
                       = *€18,422.31/yr*

pf = cos(theta) = 0.65
theta = 0.86321189 radians
kVAR initial = 1000 kW × tan (0. 86)
              = 1169 kVAR

pf = cos(theta) = 0.8
theta = 0.643501109 radians
kVAR initial = 1000 kW × tan (0.86)
               750 kVAR

capacitor size needed = *419 kVAR*

Also, using a pf correction table for 0.65 => 0.80:

kVAR = (0.419) × (1000 kW)
      = *419 kVAR*

Problem: How much could they save by owning their own transformers and switching to service level 1?

Given: A company has contacted you regarding their rate schedule. They are on the rate schedule shown in Figure 3-10, service level 5 (secondary service), but are near transmission lines and so can accept service at a higher level (service level 1) if they buy their own transformers. Assume they consume 300,000 kWh/month and are billed for 1,000 kW each month. Ignore any charges other than demand and energy.

Solution:

Service level 1 (proposed)
Demand Charge

On-peak	€10.59 /kW/mo	June-Oct.	5 months/year
Off-peak	€3.84 /kW/mo	Nov.-May	7 months/year

Energy Charge

For first two million kWh	€0.03257 /kWh
All kWh over two million	€0.02915 /kWh

Service level 5 (present)
Demand Charge

On-peak	€13.27 /kW/mo	June-Oct.	5 months/year
Off-peak	€4.82 /kW/mo	Nov.-May	7 months/year

Energy Charge

For first two million kWh	€0.03528 /kWh
All kWh over two million	€0.03113 /kWh

Rate savings:
Demand Charge

On-peak	€2.68 /kW/mo	June-Oct.	5 months/year
Off-peak	€0.98 /kW/mo	Nov.-May	7 months/year

Energy Charge

For first two million kWh	€0.00271 /kWh
All kWh over two million	€0.00198 /kWh

$$ES = 300,000 \text{ kWh/mo} \times 12 \text{ mo/yr} \times €0.00271/\text{kWh}$$
$$= €9,756 \text{ /yr}$$
$$DS = 1,000 \text{ kW } (€2.68/\text{kW/mo} \times 5 \text{ mo/yr} + €0.98/\text{kW/mo} \times 7 \text{ mo/yr})$$
$$= €20,260/\text{yr}$$
$$TS = \textbf{\textit{€30,016/yr}}$$

Problem: What is the savings from switching from priority 3 to priority 4 rate schedule?

Given: In working with a brick manufacturer, you find for gas billing that they were placed on an industrial (priority 3) schedule (see Figure 3-12) some time ago. Business and inventories are such that they could switch to a priority 4 schedule without many problems. They consume 200,000 GJ of gas per month for process needs and essentially none for heating.

Solution:

Priority 3 (present)

	Schedule	Rate	Monthly Cost (for 200,000 GJ/mo)
First	100 MJ	€19.04	€19.04
Next	2.9 GJ/mo	€5.49 /Mcf	€15.92
Next	7 GJ/mo	€5.386 /Mcf	€37.70
Next	90 GJ/mo	€4.372 /Mcf	€393.48
Next	100 GJ/mo	€4.127 /Mcf	€412.70
Next	7,800 GJ/mo	€3.445 /Mcf	€26,871.00
Over	8,000 GJ/mo	€3.399 /Mcf	€652,608.00

Total present monthly cost: €680,357.84

Total present annual cost: €8,164,294.12

Priority 4 (proposed)

	Schedule	Rate	Monthly Cost (for 200,000 GJ/mo)
First 4,000 GJ/mo or fraction thereof		€12,814	€12,814.00
Next 4000 GJ/mo		€3.168/GJ	€12,672.00
Over 8000 GJ/mo		€3.122/GJ	€599,424.00

Total present monthly cost: €624,910.00

Total present annual cost: €7,498,920.00
Annual savings from switching: *€665,374.12*

Additional Considerations

What if there exists a 20% probability that switching to the proposed rate schedule will disrupt production one more time a year for an hour?

Problem: Calculate the January electric bill for this customer.

Given: A customer has a January consumption of 140,000 kWh, a peak 15-minute demand during January of 500 kW, and a power factor of 80%, under the electrical schedule of the example in Section 3.6.
Assume that the fuel adjustment is:
€0.01/kWh

Solution:

			Quantity		Cost
Customer charge	€21.00	/mo	1	mo	€21
Energy charge	€0.04	/kWh	140,000	kWh	€5,600
Demand charge	€6.50	/kW/mo	500	kW	€3,250
Taxes	8%				
Fuel Adjustment	€0.01	/kWh	140,000	kWh	€1,400
					————
				sub-total	€10,271
					————
				tax	€822
					————
				total	***€11,093***

Problem: Compare the following residential time-of-use electric rate with the rate shown in Figure 3-6.

Given: Customer charge   €8.22   /mo
             Energy charge   €0.1230 /kWh   on-peak
                                  €0.0489 /kWh   off-peak

This rate charges less for electricity used during off-peak hours—about 80% of the hours in a year—than it does for electricity used during on-peak hours.

Solution: Each of the rates has a different on-peak period. However, if we assume that no matter which rate schedule is used that 80% of the energy is used off-peak, then average cost per kWh can be calculated as follows:

AC − (Off peak percentage of energy use)(Off-peak energy cost) +
       (1 − off-peak percentage of energy use)(On-peak energy cost)

Therefore, the average cost per kWh with the above schedule is:

AC = (80%)(€0.0489/kWh)
       + (1-80%)(€0.123/kWh)
    = *€0.06372/kWh*

And the average cost per kWh with the schedule in figure 3-6 is:

AC = (80%)(€0.0058/kWh)
       + (1-80%)(€0,10857/kWh)
    = *€0.02635/kWh*

Problem: What is the power factor of the combined load?

If they added a second motor that was identical to the one they are presently using, what would their power factor be?

Given: A small facility has 20 kW of incandescent lights and a 25-kW motor load that has a power factor of 80%.

Solution: The lamp:

$$\overline{\qquad\qquad\qquad}$$
20 kW

The motor:

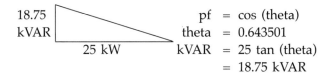

18.75 kVAR   pf = cos (theta)
                                          theta = 0.643501
                 25 kW   kVAR = 25 tan (theta)
                                                = 18.75 kVAR

Combined:

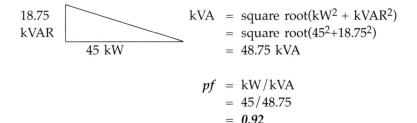

18.75 kVAR   kVA = square root(kW^2 + kVAR^2)
                                        = square root(45^2+18.75^2)
                 45 kW   = 48.75 kVA

                                   pf = kW/kVA
                                        = 45/48.75
                                        = *0.92*

Combined:

37.5 kVAR   kVA = square root(kW^2 + kVAR^2)
                                        = square root(70^2 +37.5^2)
                 70 kW   = 79.41 kVA

                                   pf = kW/kVA
                                        = 70/79.41
                                        = *0.88*

Problem: For the load curve shown below for Jones Industries, what is their billing demand and how many kWh did they use in that period?

Given: A utility charges for demand based on a 30-minute synchronous averaging period.

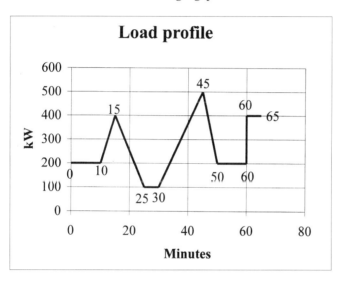

Solution: For the first 30 minutes: For the second 30 minutes:

Time (minutes)	average kW	Time (minutes)	average kW
10	200	15	300
5	300	5	350
10	250	10	200
5	100		

Weighted average: 216.67 kW   Weighted average: *275.00 kW*

Therefore, 275 kW is the billed demand.

$$kWh = (216.67 \text{ kW})(0.5 \text{ hours}) + (275 \text{ kW})(0.5 \text{ hours})$$
$$+ (400 \text{ kW})(5 \text{ minutes} \times 1 \text{ hour}/60 \text{ minutes})$$
$$= \textit{279.17 kWh}$$

Problem: Based on the hypothetical steam rate in Figure 3-13, determine their steam consumption cost for the month?

Given: The Al Best Company has a steam demand of 3,000 kg/hr and a consumption of 160,000 kg during the month of January.

Solution: <u>Steam consumption charge</u>

€3.50 /1000 kg for the first 100, 000 lb of steam per month
€3.00 /1000 kg for the next 400,000 lb of steam per month
€2.75 /1000 kg for the next 500,000 lb of steam per month
€2.00 /1000 kg for the next 1,000,000 lb of steam per month

$$
\begin{aligned}
\textit{Consumption cost} \ &= \ €3.50/1{,}000 \text{ kg} \times 100{,}000 \text{ kg} \\
&\quad + €3.00/1{,}000 \text{ kg} \times 60{,}000 \text{ kg} \\
&= \ €350 + €180 \\
&= \ \textbf{€530}
\end{aligned}
$$

Problem: What is Al's cost for chilled water in July?

What was their kJ/h equivalent for the average chilled water demand?

Given: Al Best also purchases chilled water with the rate schedule of figure 3-13. During the month of July, their chilled water demand was

485 kW and their consumption was

250,000 kWh

Solution: Chilled water demand charge:

€2,500 /mo for the first	100 kW	or any portion thereof
€15 /mo/kW for the next	400 kW	
€12 /mo/kW for the next	500 kW	
€10 /mo/kW for the next	500 kW	
€9 /mo/kW for over	1500 kW	

Chilled water consumption charge:

€0.069 /kWh	for the first 10,000 kWh/mo
€0.060 /kWh	for the next 40,000 kWh/mo
€0.055 /kWh	for the next 50,000 kWh/mo
€0.053 /kWh	for the next 100,000 kWh/mo
€0.051 /kWh	for the next 100,000 kWh/mo
€0.049 /kWh	for the next 200,000 kWh/mo
€0.046 /kWh	for the next 500,000 kWh/mo

$$\text{Demand cost} = €2,500 + (€15/kW)(385 \text{ kW})$$
$$= €8,275$$

Consumption cost =

€0.069	×	10,000	kWh/mo +
€0.060	×	40,000	kWh/mo +
€0.055	×	50,000	kWh/mo +
€0.053	×	100,000	kWh/mo +
€0.051	×	50,000	kWh/mo

= €13,690   250,000

Total bill   **= €21,965**

Average demand = consumption × 3,600 kWh/ton 744 hr/July

= 250,000 kWh × 3,600 kWh/ton 744 hr/July

= *1,209,677 kJ/h in July*

Chapter 4

Economic Analysis and Life Cycle Costing

Problem: *How* much can they spend on the purchase price for this project and still have a Simple Payback Period (SPP) of two years?

Using this figure as a cost, what is the return on investment (ROI), and the benefit-cost ratio (BCR)?

Given: The Orange and Blue Plastics Company is considering an energy management investment which will save 2,500 kWh of electric energy at €0.08/kWh. Maintenance will cost €50 per year, and the company's discount rate is 12%.

Solution: Annual savings = annual kWh saved × electric energy cost − maintenance cost
= 2,500 kWh/yr × €0.08/kWh − €50/yr
= €150/yr

Implementation cost = SPP × Annual savings
= 2 yrs × €150/yr
= *€300*

> Since no life is given, assume the project continues forever. Therefore, use the highest n in the TMV tables: n = 360

$$P = A[P|A, i, N]$$
$$300 = 150\ [P|A, i, 360]$$
$$2 = [P|A, i, 360]$$
From the TMV tables, we see that i = 50%. Therefore,
ROI = 50%

$$
\begin{aligned}
BCR &= PV(benefits)/PV\ (costs) \\
PV\ (benefits) &= A[P\,|\,A,\ i,\ N] \\
&= €150 \times [P\,|\,A,\ 12\%,\ 360\ yr] \\
&= €150 \times 8.3333 \\
&= €1{,}250 \\
BCR &= €1{,}250/€300 \\
&= \textbf{4.17}
\end{aligned}
$$

If N = 5 years:

we read from the TMV tables that the factor 2 falls between 40% and 50% tables with the factors 2.0352 and 1.7366 respectively. Therefore, to find a more precise percentage we linearly interpolate:

$$(50\% - 40\%)/(1.7366 - 2.0352)\ (X - 40\%)/(2 - 2.0352)$$

Solving for X:

$$X = \textbf{41.2 \%} = \textbf{ROI}$$

$$
\begin{aligned}
BCR &= PV\ (benefits)/PV\ (costs) \\
PV\ (benefits) &= A[P\,|\,A,\ i,\ N] \\
&= €150\ [P\,|\,A,\ 12\%,\ 5\ yr] \\
&= €150 \times 3.6048 \\
&= €541 \\
BCR &= €541/€300 \\
&= \textbf{1.80}
\end{aligned}
$$

Problem: Which model should she buy to have the lowest total monthly payment including the loan and the utility bill?

Given: A new employee has just started to work for Orange and Blue Plastics, and she is debating whether to purchase a manufactured home or rent an apartment. After looking at apartments and manufactured homes, she decides to buy one of the manufactured homes. The standard model is the basic model that costs €20,000 and has insulation and appliances that have an expected utility cost of €150 per month. The deluxe model is the energy efficient model that has more insulation and better appliances, and it costs €22,000. However, the deluxe model has expected utility costs of only €120/month. She can get a 10-year loan for 10% for the entire amount of either home.

Solution: Assume the 10% is the compounded annual percentage rate.

$$
\begin{aligned}
P &= A\,[P\,|\,A,\, i,\, N] \\
  &= A\,[P\,|\,A,\, 10\%,\, 10\text{ years}] \\
  &= A\,(6.1446) \\
A\,(standard) &= €20{,}000/6.1446 \text{ yrs} \\
  &= €3{,}255/\text{yr} \\
  &= €271/\text{mo} \\
Monthly\ (standard) &= €271/\text{mo} + €150/\text{mo} \\
  &= \overline{€421.24/\text{mo}} \\
A\,(standard) &= €22{,}000/6.1446 \text{ yrs} \\
  &= €3{,}580/\text{yr} \\
  &= €298/\text{mo} \\
Monthly\ (deluxe) &= €298/\text{mo} \quad +€120/\text{mo} \\
  &= \overline{€418.36/\text{mo}}
\end{aligned}
$$

Therefore, if she buys the deluxe, she will have a slightly lower monthly cost.

Problem: Determine the SPP, ROI, and BCR for this project:

Given: The Al Best Company uses a 7.5-kW motor for 16 hours per day, 5 days per week, 50 weeks per year in its flexible work cell. This motor is 85% efficient, and it is near the end of its useful life. The company is considering buying a new high efficiency motor (91% efficient) to replace the old one instead of buying a standard efficiency motor (86.4% efficient). The high efficiency motor cost €70 more than the standard model, and should have a 15-year life. The company pays €7 per kW per month and €0.06 per kWh. The company has set a discount rate of 10% for their use in comparing projects.

Solution: Assume the load factor (lf) is 60%.

$$DR = lf \times Pm \times ((1/effs) - (1/effh))$$

where,

DR = Demand reduction
Pm = Power rating of the motor, 7.5 kW
$effs$ = Efficiency of the standard efficiency motor, 86.4%
$effh$ = Efficiency of the high efficiency motor, 91%

Therefore,

$$DR = 0.6 \times 7.5 \text{ kW/hp} \times ((1/0.864) - (1/0.91))$$
$$= 0.26 \text{ kW}$$

$$DCR = DR \times DC \times 12 \text{ mo/yr}$$

where,

DCR = Demand cost reduction
DC = Demand cost, €7/kW/mo

Therefore,

$$DCR = 0.26 \text{ kW} \times €7/\text{kW/mo} \times 12 \text{ mo/yr}$$
$$= €22.12/\text{yr}$$

$$ES = DR \times 16 \text{ hr/day} \times 5 \text{ days/wk} \times 50 \text{ wk/yr}$$

Therefore,

$$ES = 0.26 \text{ kW} \times 16 \text{ hr/day} \times 5 \text{ days/wk} \times 50 \text{ wk/yr}$$
$$= 1,053.1 \text{ kWh/yr}$$

$$ECS = ES \times EC$$
where,
$$ECS = \text{Energy cost savings}$$
$$EC = \text{Energy cost, €0.06/kWh}$$

Therefore,
$$ECS = 1{,}047.5 \text{ kWh/yr} \times €0.06/\text{kWh}$$
$$= €63.19/\text{yr}$$

Therefore, the annual cost savings (ACS) can be calculated as follows:
$$ACS = DCS + ECS$$
$$= €22.12/\text{yr} + €63.19/\text{yr}$$
$$= €85.30/\text{yr}$$

$$SPP = \text{Cost premium ACS}$$
$$= €70.00/€85.30/\text{yr}$$
$$= \textit{0. 821 yrs}$$

Additionally, the ROI can be found with looking up the following factor in the interest rate tables:

$$P = A\ [P\,|\,A,i,N]$$
$$€70 = €84.85/\text{yr}\ [P\,|\,A,\ ROI,\ 14 \text{ years}]$$
$$0.\ 825 = [P\,|\,A,\ ROI,\ 14 \text{ years}]$$
$$\textit{ROI} = \textit{121.2\%}$$

$$BCR = PV(\text{benefits})/PV\ (\text{costs})$$
$$PV\ (\text{benefits}) = A[P\,|\,A,\ i,\ N]$$
$$= €85.85/\text{yr}\ [P\,|\,A,\ 10\%,\ 14 \text{ yr}]$$
$$= €84.85 \times 7.3667$$
$$= €628.40$$
$$BCR = €625/€70$$
$$= \textit{8.98}$$

Problem: Using the BCR measure, which project should the company select? Is the answer the same if life cycle costs (LCC) are used to compare the projects?

Given: Craft Precision, Incorporated, must repair their main air conditioning system, and they are considering two alternatives.

(1) purchase a new compressor for €20,000 that will have a future salvage value of €2,000 at the end of its 15-year life; or

(2) purchase two high efficiency heat pumps for €28,000 that will have a future salvage value of €3,000 at the end of their 15-year useful life.

The new compressor will save the company €6,500 per year in electricity costs, and the heat pumps will save €8,500 per year. The company's discount rate is 12%.

Solution:

$$BCR\ (1) = PV(\text{benefits})/PV\ (\text{costs})$$
$$PV\ (\text{benefits } 1) = A[P\,|\,A,\ i,\ N]$$
$$= €6,500/yr\ [P\,|\,A,\ 12\%,\ 15\ yr] +$$
$$€2,000[P\,|\,F,\ 12\%,\ 15]\ yr$$
$$= €6,500 \times 6.8109 + €2,000 \times 0.1827$$
$$= €44,636.25$$
$$\textbf{BCR (1)} = €44,636.25/€20,000$$
$$= \textbf{2.23}$$

$$BCR\ (2) = PV(\text{benefits})/PV\ (\text{costs})$$
$$PV\ (\text{benefits } 2) = A[P\,|\,A,\ i,\ N]$$
$$= €8,500/yr\ [P\,|\,A,\ 12\%,\ 15\ yr] + €3,000[P\,|\,F,$$
$$12\%,\ 15]\ yr$$
$$= €8,500 \times 6.8109 + €3,000 \times 0.1827$$
$$= €58,440.75$$
$$\textbf{BCR (2)} = €58,440.75/€28,000$$
$$= \textbf{2.09}$$

Therefore, since BCR(1) > BCR(2), select option 1: the new compressor.

$$LCC\ (1)\ =\ \text{Purchase cost} - \text{PV (benefits 1)}$$
$$=\ €20{,}000 - €44{,}636.25$$
$$=\ €(24{,}636.25)$$

$$LCC\ (2)\ =\ \text{Purchase cost} - \text{PV (benefits 2)}$$
$$=\ €28{,}000 - €58{,}440.75$$
$$=\ €(30{,}440.75)$$

Therefore, the answer with the LCC is different. Since LCC (2) is more negative (less cost), select option 2: the two high efficiency heat pumps.

Additional Learning Point

Why the difference?

While the BCR and NPV methods will provide the same accept or reject decisions on independent projects, these different methods may yield different rank orders of projects profitabilities for mutual exclusive projects. The difference is that the BCR method is a measure of how much each dollar invested earns. However, it does not take into account the overall size of the project. Therefore, to make a decision on which mutually exclusive project to select, one needs to use a NPV method, which takes into account the size (amount invested) of the project.

Problem: There are a number of energy-related problems that can be solved using the principles of economic analysis. Apply your knowledge of these economic principles to answer the following questions.

Given: a) Estimates of our use of coal have been made that say we have a 500 years' supply at our present consumption rate. How long will this supply of coal last if we increase our consumption at a rate of 7% per year? Why don't we need to know what our present consumption is to solve this problem?

b) Some energy economists have said that it is not very important to have an extremely accurate value for the supply of a particular energy source. What can you say to support this view?

c) A community has a 100 MW electric power plant, and their use of electricity is growing at a rate of 10% per year. When will they need a second 100 MW plant? If a new power plant costs €1 million per MW, how much money (in today's dollars) must the community spend on building new power plants over the next 35 years?

Solution::

a) Year	(yrs)	Present Use	Remaining (yrs)	Year	Present (yrs)	Use	Remaining (yrs)
0	500	1.00	500.00	27	500	6.21	420.30
1	500	1.07	498-93	28	500	6.65	413.65
2	500	1.14	497.79	29	500	7.11	406.54
3	500	1.23	496.56	30	500	7.61	398.93
4	500	1.31	495.25	31	500	8.15	390.78
5	500	1.40	493.85	32	500	812	382.07
6	500	1.50	492.35	33	500	9.33	372.74
7	500	1.61	490.74	34	500	9.98	362.76
9	500	1.72	489.02	35	500	10.68	352.09
9	500	1.84	487.18	36	500	11.42	340.66
10	500	1.97	485.22	37	500	12.22	328.44
11	500	2.10	493.11	38	500	13.08	315.36
12	500	2.25	480.86	39	500	13.99	301.36

(Continued)

a) Year	(yrs)	Present Use	Remaining (yrs)	Year	Present (yrs)	Use	Remaining (yrs)
13	500	2.41	478.45	40	500	14.97	296.39
14	500	2.58	475.87	41	500	16.02	270.37
15	500	2.76	473.11	42	500	17.14	253.22
16	500	2.95	470.16	43	500	18.34	234.88
17	500	3.16	467.00	44	500	19.63	215.25
18	500	3.38	463.62	45	500	21.00	194.25
19	500	3.62	460.00	46	500	22.47	171.79
20	500	3.87	456.13	47	500	24.05	147.73
21	500	4.14	451.99	48	500	25.73	122.00
22	500	4.43	447.56	49	500	27.53	94.47
23	500	4.74	442.82	50	500	29.46	65.01
24	500	5.07	437.75	51	500	31.52	33.50
25	500	5.43	432.32	52	500	33.73	(0.23)
26	500	5.81	426.52				

Therefore, with a present amount of coal of 500 years at the present use will only last 52 years if the use is increased by 7% a year. We do not need to know our present consumption, since we can state the consumption in terms of years.

b) New technologies will allow more efficient use of these resources. Additionally, new technologies will allow for more of these resources to be found. Furthermore, technological development will find new energy sources.

c) Assume that the present peak utilization of the power plant is 50%. Therefore, one can calculate when a new power plant is needed as follows:

yr	Present peak use (MW)	yr	Present peak use(MW)	yr	Present peak use (MW)	yr	Present peak use (MW)
0	50	9	118	18	278	27	655
1	55	10	130	19	306	28	721
2	61	11	143	20	336	29	793
3	67	12	157	21	370	30	872
4	73	13	173	22	407	31	960
5	81	14	190	23	448	32	1056
6	89	15	209	24	492	33	1161
7	97	16	230	25	542	34	1277
8	107	17	253	26	596	35	1405

Therefore, they need a new plant in year 7.

Therefore, they will need to build fourteen 100 MW power plants over the next 35 years. Assuming that they build the plants in 100 MW increments, a MARR of 10% and that the cash flow for building the plant all occurs in the year before they reach the next 100 MW increment (unlikely), then the present value of these plants can be calculated as follows:

yr	number of plants	cost (€million)	PV (€million)
7	1	100	51.32
14	1	100	26.33
18	1	100	17.99
21	1	100	13.51
24	1	100	10.15
26	1	100	8.39
27	1	100	7.63
29	1	100	6.30
30	1	100	5.73
31	1	100	5.21
32	1	100	4.74
33	1	100	4.31
34	2	200	7.83
			———
			€169.43

Therefore, these plants will cost about €169 million in today's euros.

Problem: How many hours per week must the gymnasium be used in order to justify the cost difference of a one-year payback?

Given: A church has a gymnasium with sixteen 500 Watt incandescent ceiling lights. An equivalent amount of light could be produced by sixteen 250 Watt PAR (parabolic aluminized reflector) ceiling lamps. The difference in price is €10.50 per lamp, with no difference in labor. The gymnasium is used 9 months each year. Assume that the rate schedule used is that of Problem 3.8, that gymnasium lights do contribute to the peak demand (which averages 400 kW), and that the church consumes enough electricity that much of the bill comes from the lowest cost block in the table.

Solution: Customer charge: €8.22/mo

   Energy charge   €0.1230/KWh   on-peak
                    €0.0489/kWh   off-peak

   This rate charges less for electricity used during off-peak hours—about 80% of the hours in a year—than it does for electricity used during on-peak hours.

   AC = (Off-peak percentage of energy use) (Off-peak energy cost) + (1 – off-peak percentage of energy use)(On-peak energy cost)

   Therefore, the average cost per kWh with the above schedule is:

   AC = (80%)(€0.0489/kWh)
        + (1 – 80%)(€0.123/kWh)
        = **€0.06372/kWh**

   The demand reduction (DR) from the retrofit can be calculated as follows:

   DR = N × (Do – Dnew)

where,

N = Number of lamps, 16 lamps

Do = Initial demand per lamp, 500 W/lamp

$Dnew$ = Demand of the PARs per lamp, 250 W/lamp

Therefore,

DR = 16 lamps × (500 W/lamp – 250 W/lamp)

= 4,000 W

= 4 kW

The implementation cost (IC) can be calculated as follows:

IC = Cost premium × N

= €10.50 × 16 lamps

= €168.00

SPP = IC/CS

where,

CS = Cost savings

Therefore

CS = IC/SPP

= €168.00/1 yr

= €168.00/yr

The number of weeks (Nw) the gym is used each year can be estimated as follows:

Nw = 52 wks/yr × 9 months/12 months

= 39 wks/yr

The number of hours a week (h) the lights must operate can be calculated as follows:

h = CS x DR/(Nw x Ac)

Therefore,

h = €168/yr × 4 kW/(39 wks/yr × €0.06372/kWh)

= 16.9 h/wk

Problem: Find the equivalent present worth and IRR of the following 6-year project:

Given: Use the depreciation schedule in Table 4-1
            purchase and installation cost: €100,000
            annual maintenance cost: €10,000
            annual energy cost savings: €45,000
            salvage value: €20,000
            MARR: 12%
            Tax rate: 34%
            equipment life: 5 years for depreciation purposes

Solution: assuming end of year convention

Year	Before tax cash flow	Depreciation	Taxes	After tax cash flow	PV
0	€(100,000)				€(100,000)
1	€35,000	€20,000	€5,100	€29,900	€31,250
2	€35,000	€32,000	€1,020	€33,980	€27,902
3	€35,000	€19,200	€5,372	€29,628	€24,912
4	€35,000	€11,520	€7,983	€27,017	€22,243
5	€35,000	€11,520	€7,983	€27,017	€19,860
6	€55,000	€5,760	€16,742	€38,258	€27,865

NPV: *€54,032*

The MARR that drives the present value to zero is *28.4875%*, which is the *IRR* or ROR.

Problem: Calculate the constant Euro, after tax ROR or IRR for Problem 4-7 if the inflation rate is 6%

Given: Use the depreciation schedule in Table 4-1

purchase and installation cost:	€100,000
annual maintenance cost:	€10,000
annual energy cost savings:	€45,000
salvage value:	€20,000
MARR:	12%
Tax rate:	34%
equipment life:	5 years for depreciation purposes

Solution: assuming end of year convention

Year	Before tax cash flow	Depreciation	Taxes	After tax cash flow	PV
0	(€100,000)				(€100,000)
1	€37,100	€20,000	€5,814	€31,286	€33,125
2	€39,326	€32,000	€2,491	€36,835	€31,350
3	€41,686	€19,200	€7,645	€34,040	€29,671
4	€44,187	€11,520	€11,107	€33,080	€28,081
5	€46,838	€11,520	€12,008	€34,830	€26,577
6	€69,648	€5,760	€21,722	€47,926	€35,286
				NPV:	**€84,091**

The MARR that drives the present value to zero is **21%**, which is the **IRR** or ROR.

Problem: What is the constant euro, after-tax **ROR or IRR** for this project?

Find the equivalent constant euro after-tax present worth of the following 6-year project using the depreciation schedule in Table 4-6:

Given:

Purchase and installation cost	€100,000
Annual maintenance (AM)	€10,000/yr
Maintenance cost inflation	5%/yr
Annual energy savings (ES)	€45,000
ES growth	8%/yr
Salvage value (SV)	€20,000
Salvage value growth	6%/yr
Consumer price index (CPI) growth	6%/yr
MARR in constant euros	12%/yr
Tax rate	34%/yr
Depreciation life (N)	5 yrs

Solution:

Year	PV	Constant Dollar	After tax cash flow	Taxes	Taxable income	Deprecia-ticn	Cash flow	AM	ES	SV
0	€(100,000)	€(100,000)	€(100,000)							
1	€25,185	€28,208	29,900	5,100	15,000	20,200	35,000	10,000	45,000	
2	€25,560	€32,063	36,026	2,074	6,100	32,000	38,100	10,500	48,600	
3	€20,256	€28,458	33,894	7,569	22,263	19,200	41,463	11,025	52,488	
4	€16,959	€26,686	33,690	11,421	33,591	11,520	45,111	11,576	56,687	
5	€15,392	€27,126	36,301	12,766	37,547	11,520	49,067	12,155	61,222	
6	€19,964	€39,406	55,898	25,829	75,967	5,760	81,727	12,763	66,120	28,370
	€23,317									

The MARR that drives the present value to zero is **19.69%**, which is the **IRR** or ROR

Chapter 5

Electrical Distribution Systems

5.1
Solution:
      PF (Cos Phi) = 1.0 or 100% (since it is an AC resistive load)
$$P_{IN} = V \times A \times PF \text{ (Cos Phi)}$$
$$= 220 \times 8 \times 1.0$$
$$= 1760W$$
$$= 1.76kW$$

5.2 The larger wire has a lower resistance, and the I^2R loss in the distribution system will be lower. This reduces their cost per kWh.

5.3 Single phase AC load
$$P_{IN} = V \times A \times PF \text{ (Cos Phi) W}$$
$$= kV \times A \times PF \text{ (Cos Phi) kW}$$
$$kW_{IN} = (0.24)(20)(0.8)$$
$$= 3.84 \text{ kW}$$
            Cos Phi

5.4 Electrical load is 200 kVA and 100 kW. Find PF (Cos Phi) and kVAR

$$PF = \frac{kW}{kVA} = \frac{100}{200} = .5 \text{ or } 50\%$$

$$kVAR^2 = kVA^2 - kW^2 = 200^2 - 100^2$$
$$= 30,000$$
$$kVAR = \sqrt{30,000} = 173.2 \text{ kVAR}$$

      Cos Phi

5.5   Inductive kVAR creates the electromagnetic fields that produce the torque to start and turn the motor. kVAR does not use energy. The kW the motor draws provides the real work.

5.6   380 Volts   Find PF (Cos Phi) = .72
Solution:

$$kVA = \sqrt{3} \times kV \times A = 1.732 \times .38 \times 50$$
$$= 32.91 \ kVA$$

$$kW = kVA \times PF \ (Cos \ Phi)$$
$$= 32.91 \times 0.72 = 23.7 \ kW$$

5.7   60 kW AC motor   PF (Cos Phi)
Solution:

$$kW_{IN} = \frac{60 \ kW \ | \ LF = 1}{.915}$$

$$= 65.6 \ kW$$

5.8   380 V   A = 40 P_{IN} = 24 kW
Find the PF (Cos Phi)
Solution:

$$kW_{IN} = \sqrt{3} \times kV \times A \times PF \ (Cos \ Phi)$$
$$24 = 1.732 \times .38 \times 40 \times PF \ (Cos \ Phi)$$
$$24 = 26.33 \times PF \ (Cos \ Phi)$$
$$PF \ (Cos \ Phi) = 24/26.33 = 91.2\%$$

5.9   AC induction motor draws 150 kW, and has a PF (Cos Phi) of 85%. Find the kVARs of capacitance to put on the motor to increase the PF (Cos Phi) to 90%.

kVAR = 150 x PF (Cos Phi) table number to go from 85 to 90%
    Table number = 0.136
    kVAR = 150 x 0.136 = 20.4 kVAR

5.10   a) Monthly load factor

$$MELF = \frac{350,000 \ kWh}{600 \ kW \ per} \times 720 \ hours = 81\%$$

b)   Average cost per kWh
       Energy cost   = 350,000 kWh x €0.12/kWh
                     = €42,000

$$kW_{IN} = \frac{\begin{array}{c|c|c} 600 \text{ kW} & €8 & 1 \text{ Mo} \end{array}}{\begin{array}{c|c} & kW \text{ Mo} & \end{array}}$$

              = €4,800
Total cost   = €42,000 + €4,800
              = €46,800
Average cost
per kWh   = €46,800/350,000 kWh
              = €0.1337/kWh

c) If the monthly load factor had been 25%, for othe same kWh
   find the peak kW for the month MELF = 0.25

$$= \frac{350,000 \text{ kWh}}{kW_{PEAK} \times 720 \text{ h}}$$

$$kW_{PEAK} = \frac{350,000 \text{ kWh}}{0.25 \times 720} = 1944.4 \text{ kW}$$

Find average kWh cost.
Energy cost still €42,000
Demand cost   = 1044.4 kW x €8/kW = €15,555.20

Total cost   = €15,555.20
                   €42,000.00
                   _____

                   €57,555,20

Average cost per kWh = €57,555,20/350,000 kWh
                      = €0.1644/kWh

Chapter 6

Lighting

6.1

Problem: How much can you save by installing a photocell? What is the payback period of this investment?

Given: When performing an energy survey, you find twelve two-lamp F40T12 security lighting fixtures turned on during daylight hours (averaging 12 hours/day). The lamps draw 40 Watts each, the ballasts draw 12 Watts each, and the lights are currently left on

Energy cost:	€0.055	/kWh
Power cost:	€7	/kW
Lamps:	€1	/lamp
Photocell (installed):	€85	/cell

Solution: Assuming one photocell can control all 12 fixtures

There will probably be demand savings, since the lights will be turned off during the day. It is probable that their peak demand occurs during the day. Therefore, the demand reduction (DR) can be calculated as follows:

$$DR = Nf \times N1 \times P1 + Nf \times Pb$$

where,

Nf = Number of fixtures, 12 fixtures
$N1$ = Number of lamps per fixture, 2 lamps/fixture
$P1$ = Power use of lamps, 40 W/lamp
Pb = Power use of ballasts, 12 W/fixture

Therefore,

$$DR = 12 \text{ fixtures} \times 2 \text{ lamps/fixture} \times 40 \text{ W/lamp} +$$
$$12 \text{ fixtures} \times 12 \text{ W/fixture}$$
$$= 1,104 \text{ W}$$
$$= 1.104 \text{ kW}$$

Therefore, the energy savings (ES) can be calculated as follows:

$$ES = DR \times 12 \text{ hr/day} \times 365 \text{ days/yr}$$
$$= 1.104 \text{ kW} \times 12 \text{ hr/day} \times 365 \text{ days/yr}$$
$$= 4{,}835.52 \text{ kWh/yr}$$

Therefore, the cost savings (CS) can be calculated as follows:

$$CS = ES \times €0.055/\text{kWh} +$$
$$DR \times €7/\text{kW} \times 12 \text{ mo/yr}$$
$$= \textit{€358.69/yr}$$

$$SPP = IC/CS$$
$$= €85/€358.69/\text{yr}$$
$$= 0.24 \text{ years}$$
$$= \textbf{\textit{2.84 months}}$$

6.2

Problem: What is the simple payback period (SPP) and what is the return on investment for each alternative?

Given: You count 120 four-lamp F40T12 fixtures that contain 34-Watt lamps and two ballasts. How much can you save by installing:
a. 3-F40T10 lamps at €15/fixture?
b. 3-F32T8 lamps and an electronic ballast at €40/fixture?
Assume the same energy costs as in problem 5.1.

Solution:

Energy cost:	€0.055	/kWh
Power cost:	€7	/kW
Number of fixtures:	120	fixtures
Present power use per lamp (Pp)	156.4	W/fixture including ballast
Power use per fixture for option a. (Pa)	138	W/fixture including ballast
Power use per fixture for option b. (Pb)	91.2	W/fixture including ballast
Implementation cost for option b. (ICb)	€15	/fixture
Implementation cost for option a. (ICa)	€40	/fixture
Assuming that the lights are used	876	hrs/yr
Assuming that the life of the fixtures is	7	yrs

Option	Demand Reduction (kW)	Energy Savings (kWh/yr)	Cost Savings (€/yr)	IC (€)	SPP (yrs)	IRR (%)
a.	2.21	1,934	292	1,800	*6.17*	3.3%
b	7.82	6,854	1.034	4,800	*4.64*	11.5%

6.3—Problem: How much can you save by replacing the two 20-Watt bulbs with a 7-Watt CFL?

Given: You see 25 exit signs with two 20-Watt incandescent lamps each. The 20-Watt incandescent lamps have a 2,500-hour life span and costs €3 each. The 7-Watt CFLs have a 12,000-hour life span and cost €5 each and require the use of a €15 retrofit kit. Assume the same energy costs given in Problem 5-1.

Solution:

Energy cost:	€0.055 /kWh
Power cost:	€7 /kW
Number of fixtures:	25 fixtures
Present power use per lamp (Pp)	40 W/fixture
Power use per fixture of retrofit (Pr)	7 W/fixture including ballast
Present life of lamps (Lp)	2,500 hours/lamp
Life of retrofit lamps (Lr)	12,000 hours/lamp
Present lamp cost (cp)	€3 /lamp
Retrofit lamp cost (cr)	€5 /lamp
Assuming that the lights are used	8,760 hrs/yr

Assuming that the labor needed to replace the lamp is included in lamp replacement cost

Option	Demand Reduction (kW)	Energy Savings (kWh/yr)	Energy and Demand Cost Savings (€/yr)	Annual Lamp Replacement Costs (€/yr)	Lamp Replacement Cost Savings (€/yr)	Total Annual Cost Savings (€/yr)
Present	—	—	—	525.60	—	—
Retrofit	0.825	7,227	466.79	91.25	434.35	901.14

The annual lamp replacement costs (LRC) can be calculated as follows:

LRC = Number of lamps per fixture × Number of fixtures × Replacement lamp cost × Annual lamp use/Lamp life

6.4

Problem: How can this problem be solved, and how much money can you save in the process?

Given: An old train station is converted to a community college center, and a train still passes by in the middle of the night. There are 82 75-Watt A19 lamps in surface-mounted wall fixtures surrounding the building, and they are turned on about 12 hours per day. The lamps cost €0.40 each and last for about one week before failure. Assume electricity costs 8 cents per kWh.

Solution: According to table 5-10, a replacement for a 75-Watt incandescent lamp is an 18-W compact fluorescent lamp (CFL). CFL last longer than incandescent. Additionally, according to table 5-10 this will have energy savings €34.20 over the life of the 18-W CFL. Additionally, according to table 5-5, one expects an 18-W CFL to last 10,000 hours. Since the lamps are on half the time (12-hours per day), we expect the CFL to last 2 years. Additionally, at the present time, the train station personnel replace the lamps about once a week at a cost of €0.40 per lamp or €20.80 per year (52 weeks × €0.40/wk). Additionally, we expect 18-W CFL to cost about €20 per lamp. Therefore, the cost of the replacement lamps cancel. Therefore, we can calculate the energy cost savings as follows:

$$\text{ECS} = €34.20/\text{lamp}/2 \text{ years} \times 82 \text{ lamps}$$
$$= €1,402.20/\text{yr}$$

Additionally, one would expect a labor cost savings. Assuming that the burdened labor cost is €10 per hour and the maintenance crew spends 2 man-hr/wk to replace the lamps. Therefore, the labor savings (LS) can be calculated as follows:

$$\text{LS} = €10/\text{man-hr} \times 2 \text{ man-hr}/\text{wk} \times 52 \text{ wk}/\text{yr}$$
$$= €1,040.00/\text{yr}$$

Therefore, we estimate the total annual cost savings (CS) as follows:

$$\text{CS} = \text{Lamp replacement savings} + \text{ECS} + \text{LS}$$
$$= 0 + €1,402.20/\text{yr} + 1,040/\text{yr}$$
$$= \textbf{\textit{€2,442.20/yr}}$$

6.5

Problem: How much can you save by replacing these fixtures with 70-Watt HPS cutoff luminaires?

Given: During a lighting survey you discover thirty-six 250-Watt mercury vapor cobrahead streetlights operating 4,300 hours per year on photocells.
There is no demand charge, and energy costs €0.055 per kWh.

Solution:

Energy cost:	€0.055	/kWh
Number of fixtures:	36	fixtures
Present power use per lamp (Pp)	300	W/fixture including ballast
Power use per fixture of retrofit (Pr)	84	W/fixture including ballast
Assume both lamp lives are about the same		
Assume both lamp costs are about the same		
Lamp use	4,300	hrs/yr

Option	Demand Reduction (kW)	Energy Savings (kWh/yr)	Energy and Demand Cost Savings (€/yr)
Present	—	—	—
Retrofit	7.776	33,437	**1,839.02**

6.6

Problem: What is the savings from retrofitting the facility with 250-Watt high pressure sodium (HPS) downlights?
What will happen to the lighting levels?

Given: You find a factory floor that is illuminated by eighty-four 400-Watt mercury vapor downlights. This facility operates two shifts per day for a total of 18 hours, five days per week.

Assume that the lights are contributing to the facility's peak demand, and the rates given in Problem 5-1 apply.

Solution:

Energy cost:	€0.055/kWh
Power cost:	€7/kW
Number of fixtures:	84 fixtures
Present power use per lamp (Pp)	480 W/fixture including ballast
Power use per fixture of retrofit (pr)	300 W/fixture including ballast
Assume lamp lives are about the same	
Assume lamp costs are about the same	
Assuming that the lights are used	4,680 hrs/yr

Option	Demand Reduction (kW)	Energy Savings (kWh/yr)	Energy and Demand Cost Savings (€/yr)
Retrofit	15.120	70,762	5,161.97

One would expect the lighting levels to be about the same to a little higher. Check the manufacturer's data for an exact comparison.

6.7

Problem: What will happen to the lighting levels throughout the space and directly under the fixtures? Will this retrofit be cost-effective?

What is your recommendation?

Given: An office complex has average ambient lighting levels of 27 LUX with four-lamp F40T12 40-Watt 2' × 4' recessed troffers. They receive a bid to convert each fixture to two centered F32T8 lamps with a specular reflector designed for the fixture and an electronic ballast with a ballast factor of 1.1 for €39 per fixture. This lighting is used on-peak, and electric costs are €6.50 per kW and €0.05 per kWh.

Solution: *One would expect the overall lighting levels to decrease, while the reflectors should reduce this reduction by concentrating the light to the areas below the lights.*

Energy cost:	€0.05 /kWh
Power cost:	€6.50 /kW
Number of fixtures:	1 fixture
Present power use per lamp (Pp)	184.0 W/fixture including ballast
Power use per fixture for retrofit (Pr)	70.4 W/fixture including ballast
Implementation cost for retrofit (IC)	€39 /fixture
Assuming that the lights are used	8,760 hrs/yr

Option	Demand Reduction (kW)	Energy Savings (kWh/yr)	Savings (€/yr)	Cost IC (€)	SPP (yrs)
Retrofit	0.11	995	59	39	*0.67*

Since this retrofit would pay back in less than a year, this would be a good project.

6.8

Problem: What is your advice?

Given: An exterior loading dock in Kiev, Ukraine, uses F40T12
40-Watt lamps in enclosed fixtures. They are considering a
move to use 34-Watt lamps.

Solution:

Energy cost:	€0.055	/kWh
Power cost:	€7	/kW
Number of fixtures:	1	fixtures
Present power use per lamp (Pp)	46.0	W/fixture including ballast
Power use per fixture for retrofit (Pr)	39.1	W/fixture including ballast
Implementation cost for retrofit (IC)	€1	/fixture
Assuming that the lights are used	8,760	hrs/yr

Option	Demand Reduction (kW)	Energy Savings (kWh/yr)	Cost Savings (€/yr)	IC (€)	SPP (yrs)
Retrofit	0.01	60	4	1	*0.26*

If we assume that the electric costs are the same as in Problem 5-1, the lights are on all the time, the cost of the 34-Watt
lamps is €1 more per lamp, and the lighting levels are
higher than needed (the 34-Watt lamps will produce a little
less light), then this project looks good since its payback is
less than a year.

*Therefore, my advice would be to replace the 40-Watt lamps
with 34-Watt lamps the next time they perform a group
lamp replacement.*

6.9

Problem: How would you recommend they proceed with lighting changes? What will be the savings if they have a cost of 6 cents per kWh?

Given: A turn-of-the-century power generating station uses 1500-Watt incandescent lamps in pendant mounted fixtures to achieve lighting levels of about 200 LUX in an instrument room. They plan on installing a dropped ceiling with a 60 × 120 cm grid.

Solution: There exist many strategies that could work, depending on other conditions such as the need for light at various work surfaces and the height of the ceiling. One strategy that may work is replacing each 1,500-Watt lamp with a four-lamp F32T8 fixture. This would work if the lighting level remains within acceptable levels, which could depend on how far the drop ceiling lowers the lamps towards the working surface. Perhaps, this strategy could be used in combination with task lighting. Assuming that the one four-lamp F32T8 fixture and an 18-Watt CFL for each 1,500-Watt lamp provide an acceptable lighting level, then the cost savings (CS) from this retrofit can be calculated as follows:

Energy cost:	€0.060	/kWh	
Number of fixtures:	1	fixtures	
Present power use per lamp (Pp)	1,500	W/fixture	
Power use per fixture for F32T8 (Pf)	121.6	W/fixture	including ballast
Power use per fixture for CFL (Pc)	18	W/fixture	including ballast
Assuming that the lights are used	8,760	hrs/yr	

Option	Demand Reduction (kW)	Energy Savings (kWh/yr)	Cost Savings (€/Yr)	
Retrofit	1.36	11,917	**715**	per fixture

6.10

Problem: What would be the life-cycle savings of using 13-Watt CFL in the same fixtures?

Given: A meat-packing facility uses 100-Watt A19 lamps in jarlights next to the entrance doors. These lamps cost €0.50 each and last 750 hours. The CFLs cost €15 each, and last 12,000 hours. The lights are used on-peak, and the electricity costs 8 cents per kWh.

Solution:

Energy cost:	€0.080	/kWh
Number of fixtures:	1	fixtures
Present power use per lamp (Pp)	100	W/fixture
Power use per fixture of retrofit (Pr)	13	W/fixture including ballast
Present life of lamps (Lp)	750	hours/lamp
Life of retrofit lamps (Lr)	12,000	hours/lamp
Present lamp cost (cp)	€0.500	/lamp
Retrofit lamp cost (Cr)	€15	/lamp
Assuming that the lights are used	8,760	hrs/yr
Assuming the MARR is	15%	

Assuming that the labor needed to replace the lamp is included in lamp replacement cost

Option	Demand (kW)	Energy Use (kWh/yr)	Energy and Demand Cost in one life (€/yr)	Annual Lamp Replace- ment Costs (€/yr)	PV over 12,000 hours	Total Cost Savings over 12,000 hours (€)
100-W inc	0.100	876	70.08	5.84	93.81	—
F-13-W CFL	0.013	114	9.11	10.95	24.79	69.02

6.11

Problem: What problems can you anticipate from the light trespass off the lot? How would you recommend improving the lighting? How much can you save with a better lighting source and design?

Given: A retail shop uses a 1,000-Watt mercury vapor floodlight on the corner of the building to illuminate the parking lot. Some of this light shines out into the roadway. Use the electric costs from Problem 5-7, and assume the light does not contribute to the shop's peak load.

Solution: The problems include possible liability and wasting energy by lighting an area that does not need light. One could improve the lighting design by properly aiming the light (similar to Figure 5-7) and using a more efficient light source: Table 5-10 recommends using an 880-Watt high pressure sodium (HPS).

Energy cost:	€0.050	/kWh
Number of fixtures:	1	fixtures
Present power use per lamp (Pp)	1,200	W/fixture including ballast
Power use per fixture for retrofit (Pr)	1,056	W/fixture including ballast
Assuming that the lights are used	4,380	hrs/yr

Option	Demand Reduction (kW)	Energy Savings (kWh/yr)	Cost Savings (€/yr)
Retrofit	0.144	631	€31.54

6.12

Problem: What are the energy, power, and relamping savings from using two 250-Watt HPS floodlights? What will happen to the lighting levels?

Given: A commercial pool uses four 300-Watt quartz-halogen floodlights. The lights do contribute to the facility's peak load, and the electric rates are those of Problem 5-7.

Solution:

Energy cost:	€0.050	/kWh
Demand cost:	€6.50	/kW
Present number of fixtures:	4	fixtures
Proposed number of fixtures:	2	fixtures
Present power use per lamp (Pp)	300	W/fixture including ballast
Power use per fixture for retrofit (Pr)	300	W/fixture including ballast
Assuming that the lights are used	4,380	hrs/yr

Option	Demand Reduction (kW)	Energy Savings (kWh/yr)	Cost Savings (€/yr)
Retrofit	**0.6**	**2,628**	**€178.20**

The lighting level will be increased.

6.13

Problem: What is the solution?

Given: You notice that the exterior lighting around a manufacturing plant is frequently left on during the day. You are told that this is due to safety-related issues. Timers or failed photocells would not provide lighting during dark overcast days.

Solution: The photocell sensitivity could be set to provide light even during dark overcast days. Another solution could be to provide a mixture of low sensitivity photocells and more sensitive photocells. In this way a proportion of the lights would come on during overcast days and the rest would only come on during the night or extremely dark days.

6.14

Problem: How can you solve these problems?

Given: A manufacturing facility uses F96T12HO lamps to illuminate
             the production area. Lamps are replaced as they burn out.
             These fixtures are about 15 years old and seem to have a
             high rate of lamp and ballast failure.

Solution: One could retrofit the system with a newer lighting system.
             For example, a system using T8 lamps with electronic bal-
             lasts seems appropriate. Additionally, they should imple-
             ment a group relamping program, which would eliminate
             the need to replace lamps one-by-one as they fail. Thereby,
             these two recommendations would not only save energy, but
             would also save labor costs.

Chapter 7

Motors and Drives

Problems

7.1 An AC induction motor has a full load speed of 1465 RPM. What is its no load speed?

7.2 A 20 kW AC motor operates at full load, and draws 25 kW. What is the efficiency of that motor?

7.3 A 30 kW motor is running a 15 kW pump. What is the load factor of the motor?

7.4 The motor in Problem 7.2 is a three phase, 380 volt motor, that draws 40 amps at full load. What is the full load power factor (Cos Phi) of the motor?

7.5 A 40 kW AC induction motor has a full load speed of 1460 RPM. The actual running speed was measured in a manufacturing laboratory at 1473 RPM. The manufacturing lab tells us that the motor load factor would be the ratio of the (no load speed minus the actual running speed) divided by the (no load speed minus the full load speed). What is the load factor of the motor?

7.6 A 100 nameplate kW AC induction motor has a load factor of 70%, and an efficiency of 92.3%. What is the kW input to the motor?

7.7   A three phase 380 volt AC induction motor draws a current of 80 amperes, and has a power factor (Cos Phi) of 85%. What is the kW input to the motor?

7.8   An AC induction motor draws 100 kW and has a power factor (Cos Phi) of 75%. How many kVAR of capacitance should we put on the motor to get the power factor (Cos Phi) up to 85%?

7.9   A 40 kW motor with a load factor of 75% and an efficiency of 89.3% will be replaced with a 30 kW motor with a load factor of 100% and an efficiency of 93.6%. How many kW of power savings will be obtained from this project?

7.10 A 10 kW centrifugal fan is supplying 900 L/s on a very hot day. On a cooler day, the VFD drops the fan speed down to where it supplies only 600 L/s. How many kW is the fan motor supplying for the new 600 L/s?

Solution:
7.1   1500 RPM.

7.2   20 kW motor   25 kW load
Solution:
$$\text{Eff} = \frac{\text{Out}}{\text{Input}} = \frac{20}{25} = 80\%$$

7.3   30 kW motor   15 kW pump
Solution:
$$\text{Motor LF} = \frac{15 \text{ kW}}{30 \text{ kW}} = 50\%$$

7.4 k_{IN} for motor is 25 kW
3 phase   380 V   40 A full load

Find FL   PF (Cos Phi)

Solution:

$kW_{IN} = \sqrt{3} \times kV \times A \times PF$ (Cos Phi)

$\quad\quad 25 = (1.732) \times .38 \times 40$ PF (Cos Phi)

$\quad\quad 25 = 26.16 \times PF$ (Cos Phi)

$\quad\quad PF$ (Cos Phi) $= 25/26.16 = 95.6\%$

7.5 40 kW motor   FL 1460 RPM

Solution:

Run 1473 RPM

Load   NL 1500

$$\text{Load} = \frac{1500 - 1473}{1500 - 1460} = \frac{27}{40} = 67.5\%$$

7.6 100 kW motor   LF = 0.7

$\quad\quad$ Eff = 92.3%   Find kW_{IN}

Solution:

$kW_{IN} = (100kW) \times (0.7)/.923$

$\quad\quad\quad = 75.84$ kW

7.7 380 V   80 A   PF (Cos Phi) = .85

Solution:

$kW_{IN} = \sqrt{3} \times V \times A \times PF$ (Cos Phi)

$\quad\quad\quad = 1.732 \times 0.38 \times 80 \times 0.85$

$\quad\quad\quad = 44.75$ kW

7.8 100 Load kW   PF (Cos Phi) = 0.75

Find kVAR to get PF (Cos Phi) = 0.85

Solution:

$\quad\quad$ kVAR = 100 (.262) = 26.2 kVAR

7.9 40 kW motor LF = 0.75

$\quad\quad$ EFF = .893%   New 30 kW LF = 100%   EFF = .936

Solution:

$kW_{IN_1} = 40 \times .75/.893 = 33.5946$

$kW_{IN_2} = 30 \times 1/.936 = \dfrac{32.0513}{.5433}$

$\Delta kW = .5433$ kW savings

7.10 10 kW motor
      900 L/s —> 600 L/s

Solution:

   kW_{NEW}

$$\frac{kW_{NEW}}{kW_{OLD}} = \left(\frac{LP/s_{NEW}}{LP/s_{OLD}}\right)^3$$

$$\frac{kW_{NEW}}{10\ kW} = \left(\frac{600}{900}\right)^3 = 0.296q$$

$kW_{NEW} = 2.96$ kW

Chapter 8

Heating, Ventilating, and Air Conditioning

Problems

8.1 Estimate the total heating load caused by a work force of 22 people including 6 overhead personnel, primarily sitting during the day; 4 maintenance personnel and supervisors; and 12 people doing heavy labor. Assume that everyone works the same 8-hour day.

Solution: Assuming all the people are males, the heat load (q) can be estimated as follows:

$$q = [Ns \times qs + Nn \times qn + Nh \times qh] \times h$$

where,

Ns = Number of people seated, 6 people

qs = Heat gain from seated people, 422 kJ/h/person

Ns = Number of people doing light machine work, 4 people

qs = Heat gain from people doing light machine work, 1,097 kJ/h/person

Ns = Number of people doing heavy work, 12 people

qs = Heat gain from people doing heavy work, 1,688 kJ/h/person

h = Number of working hours, 8 hrs/day

Therefore,

q = [6 people × 400 kJ/h/person + 4 people × 1,097 kJ/hr/person + 12 people × 1,688 kJ/ person] × 8 hrs/day

= *217,408 kJ/day*

= *27,176 kJ/h*

8.2

Problem: **8.2** If the HVAC system that removes the heat in Problem 8.1 has a COP of 2.0 and runs continuously, how many kW will this load contribute to the electrical peak if the peak usually occurs during the working day? Assume that the motors in the HVAC system are outside the conditioned area and do not contribute to the cooling load.

Solution:

$$EER = COP \times 3.6 \text{ kJ/Wh}$$
$$= 2 \times 3.6 \text{ kJ/Wh}$$
$$= 7.2 \text{ kJ/Wh}$$
$$W = \text{kJ/h cooling}/(EER)$$
$$= 27{,}176 \text{ kJ/h}/7.2 \text{ kJ/Wh}$$
$$= 3{,}774.9 \text{ W}$$
$$= \mathbf{3.77 \text{ } kW}$$

8.3

Problem: Answer Problem 8.2 under the assumption that 8 of the 12 people doing heavy labor and 2 foremen-maintenance personnel come to work when the others are leaving and that 3000 Watts of extra lighting are required for the night shift.

Solution: Assuming all the people are males, the heat load (q) can be estimated as follows:

$$q = [\text{Lighting load} \times 3.6 \text{ MJ/kWh} + Nn \times qn + Nh \times qh] \times h$$

where,

Ns = Number of people doing light machine work, 2 people

qs = Heat gain from people doing light machine work, 1,097 kJ/h/person

Ns = Number of people doing heavy work, 8 people

qs = Heat gain from people doing heavy work, 1,688 kJ/h/person

h = Number of working hours, 8 hrs/day

Therefore,

$$q = [3 \text{ kW} \times 3,600 \text{ kJ/kWh} + 2 \text{ people} \times 1,097 \text{ kJ/hr/person} + 8 \text{ people} \times 1,688 \text{ kJ/h/person}] \times 8 \text{ hrs/day}$$
$$= 211,984 \text{ kJ/day}$$
$$= 26,498 \text{ kJ/h}$$

$$W = \text{kJ/h cooling/(EER)}$$
$$= 26,498 \text{ kJ/h/7.200 kJ/Wh}$$
$$= 3,680.3 \text{ W}$$
$$= \mathbf{\mathit{3.68\ kW}}$$

8.4

Problem: A 30 kW air conditioning unit has an EER of 6.3. What is the COP of that unit? How many kW of input power does that unit draw when it operates at full load mechanical cooling?

30 kW AC has EER = 6.3.
What is the COP? Find kW_{INPUT}.

Solution:

$$COP = \frac{EER}{3.6 \text{ kJ/Wh}} = \frac{6.3 \text{ kJ/Wh}}{3.6 \text{ kJ/Wh}}$$

$$= 1.75$$

Find the kW_{IN} for this unit.

$$kW_{INPUT} = \frac{30 \text{ } kW_{OUTPUT}}{1.75} = 17.14 \text{ kW}$$

8.5

Problem: If Crown Jewels buys a new motor, which one of these incentives should they ask for?

Solution:

$$\frac{kW_{IN}}{kW_{COOL}} = \frac{3.6 \text{ kJ/Wh}}{13.5 \text{ kJ/Wh}} = 0.27$$

8.6

Problem: Find kWh of input to provide 1.0 million kJ of cooling.

Solution:

$$\frac{kW_{IN} \times h}{kW_{COOL} \times h} = \frac{1}{COP}$$

$$kWh_{IN} = (1/3) \times 120 \text{ million kWh}$$
$$= 40 \text{ million kWh}$$

8.7

Problem: How many kWh electric input is used to provide 100 million kJ of heat removal (air conditioning) with a system having a COP of 5.0?

Find kWh of input to provide 100 million kJ of cooling.

Solution:

100 million kJ = 100,000 MJ
1 kWh = 3.6 MJ

$$kWh_{INPUT} = \frac{100,000 \text{ MJ}}{3.6 \text{ MJ/kWh}}$$

$$= 27,778 \text{ kWh}$$

8.8

Problem: Heat 20 L/s of 20 degree C air to 32 degree air. Find the W of sensible heat.

Solution:

$$W = 1.2 \times (L/s) \times \Delta T$$
$$= 1.2 \times 20 \times (32 - 20)$$
$$= 288W$$

8.9

Problem: AC unit removes 120,000 kJ/h of heat, and has a power input of 12 kW.

Find the COP of the AC unit.

Solution:

$$12 \text{ kW} = \frac{12 \text{ kW} \mid 3600 \text{ kJ/h}}{\text{kW}} = 43,200 \text{ kJ/h}$$

$$COP = \frac{120,000 \text{ kJ/h}}{43,200 \text{ kJ/h}} = 2.78$$

8-10

Problem: 100 m^2 wall, U = 1.4W/m^2°C), and HDD = 3000°C days/ year.
Find the heat in Wh through the wall for the heating season.

Solution:

$$Q = U \times A \times HDD \times 24$$
$$= (1.4) \times (100) \times (3000) \times (24) = 10,080,000 \text{ Wh/yr}$$

Chapter 9

Combustion Processes and the Use of Industrial Wastes

Problem: What is the after-tax present worth of the first 5 years of cash flows associated with this investment if the company uses a constant-euro after-tax rate of return of 8% on this kind of investment?

Given: In Table 9.2, a waste-burning boiler was described. Assume the capacity of this boiler is 62,000 kg/h. Suppose that these figures are 5 years old, that your company is contemplating the purchase of such a boiler, and that it is planned to save twice the energy amounts and have twice the capacity of the given boiler. The energy cost has been inflating at 10% per year, base construction costs have been inflating at 6% per year, the base inflation rate of the economy is 5%, and without inflation the cost of constructing a unit is $R^{0.73}$, multiplied by the cost of the existing unit, where R is the ratio between the capacity of the proposed unit and the capacity of the present unit. The combined rate of the company is 34%. The unit is subject to the 5-year depreciation schedule shown in Table 4-6.

	2000	2005
boiler capacity:	62,000 kg/h	56,000 kg/h
energy savings:	350,000 /yr	€700,000 /yr in 2000 euros
cost of construction:	$R^{0.73}$	
R:	2	
cost of construction:	€3,540,000	€5,871,582 (€3,540,000 x $2^{0.73}$) in 2000 euros
tax rate:		34%

	2000 through 2005
energy cost inflation:	10% /yr
construction cost inflation:	6% /yr
other inflation:	5% /yr
Hurdle rate (MARR):	8%

You have the following data on the economics of a waste-burning system:

Savings (year 2000 euros)		Costs (year 2000 euros)	
Coal and natural gas:	€350,000/yr	Site preparation:	€335,000
Trash hauling		Building to house system:	€625,000
and landfill:	€473,000/yr	Equipment support structures:	€175,000
		Boiler and trash-handling	
		equipment:	€1,560,000
		Piping:	€275,000
		Instrumentation:	€220,000
		Crew locker room:	€175,000
		Miscellaneous mechanical	
		equipment:	€115,000
		Spare parts:	€60,000

Depreciation Schedule

Year	Depreciation
1	20.00%
2	32.00%
3	19.20%
4	11.52%
5	11.52%

Assume the construction takes a year; therefore, savings start a year after construction begins. Furthermore, that construction begins in 2005 with the recognition of costs and savings at the beginning of each year.

Solution:

Cost of Construction

Year	Inflation	Cost
2000	6%	€ 5,871,582.38
2001	6%	€ 6,223,877.33
2002	6%	€ 6,597,309.97
2003	6%	€ 6,993,148.57
2004	6%	€ 7,412,737.48
2005	6%	€ 7,857,501.73

Energy Savings

Year	Inflation	Savings
2000	10%	€ 350,000.00
2001	10%	€ 385,000.00
2002	10%	€ 423,500.00
2003	10%	€ 465,850.00
2004	10%	€ 512,435.00
2005	10%	€ 563,678.50
2006	10%	€ 620,046.35
2007	10%	€ 682,050.99
2008	10%	€ 750,256.08
2009	10%	€ 825,281.69
2010	10%	€ 907,809.86

Trash Hauling and Landfill Savings

Year	Inflation	Savings
2000	5%	€ 473,000.00
2001	5%	€ 496,650.00
2002	5%	€ 521,482.50
2003	5%	€ 547,556.63
2004	5%	€ 574,934.46
2005	5%	€ 603,681.18
2006	5%	€ 633,865.24
2007	5%	€ 665,558.50
2008	5%	€ 698,836.42
2009	5%	€ 733,778.25
2010	5%	€ 770,467.16

Year	Initial investment	Depreciation Cost	Energy Savings	Hauling and Landfill Savings	Before tax savings	After tax savings (before tax x (1 - 34%))	[P|A, i, N] factor (1/(1 + MARR)^yr))	PV Sub-Totals
2005	(€ 7,857,502)					(€ 7,857,502)	1	(€ 7,857,502)
2006		(€ 1,571,500)	€ 620,046	€ 633,865	(€ 317,589)	(€ 209,609)	0.92593	(€ 194,082)
2007		(€ 2,514,401)	€ 682,051	€ 665,558	(€ 1,166,791)	(€ 770,082)	0.85734	(€ 660,221)
2008		(€ 1,508,640)	€ 750,256	€ 698,836	(€ 59,548)	(€ 39,302)	0.79383	(€ 31,199)
2009		(€ 905,184)	€ 825,282	€ 733,778	€ 653,876	€ 431,558	0.73503	€ 317,208
2010		(€ 905,184)	€ 907,810	€ 770,467	€ 773,093	€ 510,241	0.68058	€ 347,262
		(€ 7,404,910)					NPV =	(€ 8,078,534)

Therefore, the present value of the first five years is €(8,074,774)

Problem   The choice of an optimum combination of boiler sizes in the garbage-coal situation is not usually easy. Suppose that health conditions limit the time garbage, even dried, can be stored to 1 month. Use initial costs given in the accompanying table, and assume the municipality and your company have supplies and needs for energy, respectively, as given in the table labeled "data" for Problem 7.2. Suppose all other costs for this problem are the same as Section 7.4.2. What is the optimum choice now?

Costs for Problem 9.2

Capacity, 750 psi kg/h	Initial Costs: Trashed-fired boiler	Initial Costs: Coal-fired boiler
22,700 kg/h	n/a	€1,800,000
45,400	n/a	€3,500,000
68,100	€6,250,000	€5,100,000
90,800	€8,640,000	€6,900,000
113,500	€10,870,000	€8,900,000
136,000	€13,000,000	€11,000,000

Month	Garbage needed (tonnes)	Garbage available (tonnes)
January	23,000	13,500
February	23,000	13,500
March	21,600	16,500
April	19,500	18,000
May	14,100	18,900
June	9,500	19,500
July	7,600	22,500
August	9,500	21,000
September	10,800	21,000
October	13,500	18,000
November	18,400	15,000
December	24,300	18,600

Assume that garbage density is 1,304 kg/m^3 and has 34.84 MJ/m^3.

Given:

Hurdle rate (MARR):	10%	
Project life:	20	years
Salvage value:	€0	
Tax rate:	0%	
Landfill cost inflation (<5yrs.):	30%	/yr
Landfill cost inflation (>5yrs.):	10%	/yr
Assume no other inflation		

Assume that the projects are expensed; therefore, depreciation is not a factor in the present value analysis.

garbage density:	1,304	kg/m^3
garbage energy content:	34.84	MJ/m^3
garbage storage constraint:	1	month
hours of operation per year:	8,760	h/yr

Assume that the capacities in the above table already account for maintenance time and outages.

Coal energy content:	21,000,000	kJ/tonne
Cost of coal:	€55.00	/tonne
Coal ash rate:	9.6%	
Trash ash rate:	16%	

Next, we figure out our shortages based on need by month.

Month	Garbage needed (tonnes)	Garbage burned (tonnes)	Two Boilers – Coal fired 22,700 kg/hr tonnes of coal (1)	Two Boilers – Coal fired 45,400 kg/hr tonnes of coal (2)	One boiler – Trash fired tonnes Garbage from other companies (3)
January	23,000	13,500	11.46	11.46	9,500
February	23,000	13,500	11.46	11.46	9,500
March	21,600	16,500	6.15	6.15	5,100
April	19,500	18,000	3.53	1.81	1,500
May	14,100	14,100	-	-	-
June	9,500	9,500	-	-	-
July	7,600	7,600	-	-	-
August	9,500	9,500	-	-	-
September	10,800	10,800	-	-	-
October	13,500	13,500	-	-	-
November	18,400	18,400	2.21	-	-
December	24,300	24,300	9.32	-	-
Annual totals:	194,800	169,200	44.12	30.87	25,600

The calculations for January for the above table are as follows:

(1) Coal needed = (garbage needed - min(garbage burned, monthly capacity) × 2,000 kg/ton ×
    (1/garbage density) × garbage energy content × 1/coal energy content
  = (23,000 tonnes - min(13,500 tonnes,18,250 tonnes) × 1,000 kg/tonne ×
    (1 m³/1,304 kg) × 34.84 MJ/m³ × 1,000 kg/Mg × tonne/21,155,000 kJ)
  = 11.46 tonnes of coal

(2) Coal needed = (garbage needed - min(garbage burned, monthly capacity) × 1,000 kg/tonne × (1/garbage density)
    × garbage energy content × 1/coal energy content
  = (23,000 tonnes - min(13,500 tonnes, 36,500 tonnes) × 1,000 kg/tonne ×
    (1 m³/1,304 kg) × 34.84 MJ/m³ × 1,000 kg/Mg × tonne/21,155,000 kJ)
  = 11.46 tonnes of coal

(3) Garbage needed other = (garbage needed - min(garbage burned, monthly capacity)
  = (23,000 tonnes - min(13,500 tonnes, 54,750 tonnes)
  = 9,500.00 tonnes of garbage from other companies

Next, we figure out the garbage that still needs to be disposed. The only option that does not have enough capacity is the coal fired 22,700 kg/hr:

Month	Garbage burned (tonnes)	Capacity (tonnes)	Two Boilers—Coal fired 22,700 kg/hr tonnes of garbage left over
January	13,500	18,250	
February	13,500	18,250	
March	16,500	18,250	
April	18,000	18,250	
May	14,100	18,250	
June	9,500	18,250	
July	7,600	18,250	
August	9,500	18,250	
September	10,800	18,250	
October	13,500	18,250	
November	18,400	18,250	150
December	24,300	18,250	6,050
Annual totals:			6,200

Next, we calculate the ash waste for each option:

Month	Two Boilers— Coal fired 22,700 kg/hr	Two Boilers— Coal fired 45,400 kg/h	One Boiler— Trash fired
tonnes of coal	44.12	30.87	0
coal ash rate	9.6%	9.6%	9.6%
tonnes of trash	163,000	169,200	194,800
trash ash rate	16%	16%	16%
annual ash (tonnes)	26,084.24	27,074.96	31,168.00
ash hauling (€1.25/T)	€32,604.29	€33,843.70	€38,960.00
ash land fill (€2.50/T)	€65,210.59	€67,687.41	€77,920.00

annual ash = tonnes of coal × coal ash rate + tonnes of ash × trash ash rate

Next, we adjust the costs from problem 7.3 for our 3 options.

	Present System	Two Boilers -- Coal fired 22,700 kg/hr	Two Boilers -- Coal fired 45,400 kg/hr	One Boiler -- Trash fired
First Cost	None	€ 1,800,000	€ 3,500,000	€ 6,250,000
Annual Cost				
Gas	€ 2,500,000	€ 0	€ 0	€ 0
		€2,426.60	€1,697.86	
Coal	€0	44.12 tonnes x €55/tonne	30.87 tonnes x €55/tonne	€0
Boiler				
Maintenance	€ 50,000	€ 300,000	€ 300,000	€ 250,000
Waste	€310,125	€7,750		
Transportation	(248,100 T x €1.25/T)	(6,200 T x €1.25 / T)	€0	€0
Waste		15,500		
Land filling	€620,250	(6,200 T x €2.50 / T)		
(first year)	(248,100 T x €2.50)		€0	€0
Ash				
Transportation	€0	€ 32,605	€ 33,844	€ 38,960
Ash				
Land filling	€0	€ 65,211	€ 67,687	€ 77,920
(first year)				
Annual Revenues				
Waste				€384,000
from other				(25,600 T x €15.00/T)
Companies				

Finally, we fill in the cash flows:

Coal 22,700 kg/hr

Year	Initial investment (€1,800,000)	Fuel Savings (€2,500,000 - €2,427.60)	Maint. Savings (€50,000 - €300,000)	Hauling Savings (€310,125 - €7,750 - €32,605)	Landfill Savings (€620,250 - €15,500 - €65,211) then adjust for landfill increases	Landfill inflation	Revenues	FV Sub-totals	[P,A,I,N] factor (1/(1 + MARR)^yr)	PV Sub-Totals
0	(€1,800,000)							(€1,800,000)	€1	(€1,800,000)
1		€2,497,573	(€250,000)	€269,770	€539,539	30%	€0	€3,056,882	€1	€2,778,984
2		€2,497,573	(€250,000)	€269,770	€701,401	30%	€0	€3,218,744	€1	€2,660,119
3		€2,497,573	(€250,000)	€269,770	€911,321	30%	€0	€3,429,164	€1	€2,576,382
4		€2,497,573	(€250,000)	€269,770	€1,185,367	30%	€0	€3,702,711	€1	€2,529,001
5		€2,497,573	(€250,000)	€269,770	€1,540,977	30%	€0	€4,058,321	€1	€2,519,898
6		€2,497,573	(€250,000)	€269,770	€1,695,075	10%	€0	€4,212,418	€1	€2,377,800
7		€2,497,573	(€250,000)	€269,770	€1,864,583	10%	€0	€4,381,926	€1	€2,248,621
8		€2,497,573	(€250,000)	€269,770	€2,051,041	10%	€0	€4,568,384	€0	€2,131,185
9		€2,497,573	(€250,000)	€269,770	€2,256,145	10%	€0	€4,773,488	€0	€2,024,425
10		€2,497,573	(€250,000)	€269,770	€2,481,759	10%	€0	€4,999,103	€0	€1,927,371
11		€2,497,573	(€250,000)	€269,770	€2,729,935	10%	€0	€5,247,279	€0	€1,839,139
12		€2,497,573	(€250,000)	€269,770	€3,002,929	10%	€0	€5,520,272	€0	€1,758,929
13		€2,497,573	(€250,000)	€269,770	€3,303,222	10%	€0	€5,820,565	€0	€1,686,010
14		€2,497,573	(€250,000)	€269,770	€3,633,544	10%	€0	€6,150,887	€0	€1,619,721
15		€2,497,573	(€250,000)	€269,770	€3,996,398	10%	€0	€6,514,242	€0	€1,559,458
16		€2,497,573	(€250,000)	€269,770	€4,396,588	10%	€0	€6,913,932	€0	€1,504,673
17		€2,497,573	(€250,000)	€269,770	€4,836,247	10%	€0	€7,353,590	€0	€1,454,869
18		€2,497,573	(€250,000)	€269,770	€5,319,372	10%	€0	€7,837,215	€0	€1,409,592
19		€2,497,573	(€250,000)	€269,770	€5,851,359	10%	€0	€8,369,202	€0	€1,368,431
20		€2,497,573	(€250,000)	€269,770	€6,437,445	10%	€0	€8,954,388	€0	€1,331,013

NPV = €37,505,621

Coal 45,400 kg/hr

Year	Initial investment (€3,500,000)	Fuel Savings (€2,500,000)	Maint. Savings (€50,000 - €250,000)	Hauling Savings (€310,125 - €33,844)	Landfill Savings (€620,250 - €67,687) then adjust for landfill increases	Landfill inflation	Revenues	FV Sub-totals	[P/A, i, N] factor (1/(1 + MARR)^yr)	PV Sub-totals
0	(€3,500,000)							(€3,500,000)	€1	(€3,500,000)
1		€2,500,000	(€200,000)	€276,281	€552,563	30%	€0	€3,128,844	€1	€2,844,404
2		€2,500,000	(€200,000)	€276,281	€718,332	30%	€0	€3,294,613	€1	€2,722,821
3		€2,500,000	(€200,000)	€276,281	€933,831	30%	€0	€3,510,112	€1	€2,637,199
4		€2,500,000	(€200,000)	€276,281	€1,213,981	30%	€0	€3,790,262	€1	€2,588,800
5		€2,500,000	(€200,000)	€276,281	€1,578,175	30%	€0	€4,154,456	€1	€2,579,590
6		€2,500,000	(€200,000)	€276,281	€1,735,993	10%	€0	€4,312,274	€1	€2,434,166
7		€2,500,000	(€200,000)	€276,281	€1,909,592	10%	€0	€4,485,873	€1	€2,301,962
8		€2,500,000	(€200,000)	€276,281	€2,100,551	10%	€0	€4,676,832	€0	€2,181,777
9		€2,500,000	(€200,000)	€276,281	€2,310,606	10%	€0	€4,886,887	€0	€2,072,517
10		€2,500,000	(€200,000)	€276,281	€2,541,667	10%	€0	€5,117,948	€0	€1,973,190
11		€2,500,000	(€200,000)	€276,281	€2,795,834	10%	€0	€5,372,115	€0	€1,882,893
12		€2,500,000	(€200,000)	€276,281	€3,075,417	10%	€0	€5,651,698	€0	€1,800,805
13		€2,500,000	(€200,000)	€276,281	€3,382,959	10%	€0	€5,959,240	€0	€1,726,179
14		€2,500,000	(€200,000)	€276,281	€3,721,255	10%	€0	€6,297,536	€0	€1,658,338
15		€2,500,000	(€200,000)	€276,281	€4,093,380	10%	€0	€6,669,661	€0	€1,596,664
16		€2,500,000	(€200,000)	€276,281	€4,502,718	10%	€0	€7,078,999	€0	€1,540,596
17		€2,500,000	(€200,000)	€276,281	€4,952,990	10%	€0	€7,529,271	€0	€1,489,626
18		€2,500,000	(€200,000)	€276,281	€5,448,289	10%	€0	€8,024,570	€0	€1,443,289
19		€2,500,000	(€200,000)	€276,281	€5,993,118	10%	€0	€8,569,399	€0	€1,401,165
20		€2,500,000	(€200,000)	€276,281	€6,592,429	10%	€0	€9,168,710	€0	€1,362,870

NPV = €36,738,854

Trash 68,100 kg/hr

Trash 68,100 kg/hr

Year	Initial investment	Fuel Savings (€2,500,000 - €1,697.98)	Maint. Savings (€50,000 - €300,000)	Hauling Savings (€310,125 - €38,960)	Landfill Savings (€620,250 - €7*,920) then adjust for landfill increases	Landfill inflation	Revenues	FV Sub-totals	[P/A, i, N] factor (1/(1 + MARR)^yr)	PV Sub-Totals
0	(€6,250,000)							(€6,250,000)	€1	(€6,250,000)
1		€2,498,302	(€250,000)	€271,165	€542,330	30%	€384,000	€3,445,797	€1	€3,132,543
2		€2,498,302	(€250,000)	€271,165	€705,029	30%	€384,000	€3,608,496	€1	€2,982,228
3		€2,498,302	(€250,000)	€271,165	€916,538	30%	€384,000	€3,820,005	€1	€2,870,026
4		€2,498,302	(€250,000)	€271,165	€1,191,499	30%	€384,000	€4,094,966	€1	€2,796,917
5		€2,498,302	(€250,000)	€271,165	€1,548,549	30%	€384,000	€4,452,416	€1	€2,764,600
6		€2,498,302	(€250,000)	€271,165	€1,703,844	10%	€384,000	€4,607,311	€1	€2,600,707
7		€2,498,302	(€250,000)	€271,165	€1,874,228	10%	€384,000	€4,777,695	€1	€2,451,713
8		€2,498,302	(€250,000)	€271,165	€2,061,651	10%	€384,000	€4,965,118	€0	€2,316,264
9		€2,498,302	(€250,000)	€271,165	€2,267,816	10%	€384,000	€5,171,283	€0	€2,193,129
10		€2,498,302	(€250,000)	€271,165	€2,494,537	10%	€384,000	€5,398,065	€0	€2,081,188
11		€2,498,302	(€250,000)	€271,165	€2,744,067	10%	€384,000	€5,647,524	€0	€1,979,423
12		€2,498,302	(€250,000)	€271,165	€3,018,443	10%	€384,000	€5,921,930	€0	€1,886,909
13		€2,498,302	(€250,000)	€271,165	€3,320,349	10%	€384,000	€6,223,776	€0	€1,802,806
14		€2,498,302	(€250,000)	€271,165	€3,652,340	10%	€384,000	€6,555,807	€0	€1,726,349
15		€2,498,302	(€250,000)	€271,165	€4,017,574	10%	€384,000	€6,921,041	€0	€1,656,842
16		€2,498,302	(€250,000)	€271,165	€4,419,371	10%	€384,000	€7,322,799	€0	€1,593,654
17		€2,498,302	(€250,000)	€271,165	€4,861,265	10%	€384,000	€7,764,732	€0	€1,536,211
18		€2,498,302	(€250,000)	€271,165	€5,347,351	10%	€384,000	€8,250,858	€0	€1,483,989
19		€2,498,302	(€250,000)	€271,165	€5,882,130	10%	€384,000	€8,785,597	€0	€1,436,515
20		€2,498,302	(€250,000)	€271,165	€6,470,343	10%	€384,000	€9,373,810	€0	€1,393,357

NPV = €36,435,371

Therefore, as far as NPV is concerned the options look very similar. Some more sensitivity analysis needs to be performed. However, the two boiler, coal / trash fired option does have the highest net present value and should be chosen assuming the sensitivity analysis holds this to be true.

Chapter 10

Steam Generation and Distribution

Problem: How much do these leaks cost per year in lost fuel?

Given: An audit of a 4,200 KpA steam distribution system shows 50 wisps (estimated at 10 kg/h), 10 moderate leaks (estimated at 50 kg/h), and 2 leaks estimated at 350 kg/h each. The boiler efficiency is 85%, the ambient temperature is 20°C, and the fuel is coal, at €65/tonne and 30 MJ/kg. The steam system operates continuously throughout the year.

Solution: The amount of steam energy lost (q) can be estimated as follows:

$$
\begin{aligned}
q \quad &= \text{h} \times \text{summation [number of leaks} \times \text{mass flow rate of leaks]} \\
&= (2,799.58 - 83) \text{ kJ/kg} \times (50 \text{ leaks} \times 10 \text{ kg/h/leak} + 10 \\
&\quad \text{leaks} \times 50 \text{ kg/h/leak} + 2 \text{ leaks} \times 350 \text{ kg/h/leak}) \\
&= 4,618,186 \quad \text{kJ/h} \\
&= 40,455 \qquad \text{GJ/yr}
\end{aligned}
$$

Therefore, the cost (C) of these leaks can be estimated as follows:

$$
\begin{aligned}
C \quad &= \quad q \times €65/\text{tonne}/30 \text{ MJ/kg}/1,000 \text{ kg/tonne}/0.85 \\
&= \quad 40,455 \text{ GJ/yr} \times \\
&\quad\quad €65/\text{tonne}/30 \text{ MJ/kg}/1,000 \text{ kg/tonne}/0.85 \\
&= \quad \textit{€ 103,121.38/yr}
\end{aligned}
$$

Problem: How much heat is exchanged per kilogram of entering steam?

Given: Steam enters a heat exchanger at 300°C and 8.5 MPa and leaves as water at 150°C and 0.5 MPa.

Solution: Look up the enthalpy on steam tables or a Mollier diagram.

You may need to extrapolate numbers for those not directly on the steam table.

$$
\begin{aligned}
\textit{delta } h &= \ \ h1 - h0 \\
&= \ \ (2750.3 - 639.8)\ kJ/kg \\
&= \ \ \textbf{2110.5}\ \ \textit{kJ/kg}
\end{aligned}
$$

Problem: What would be the potential annual savings in the example of Section 7.5 if the amount of boiler blowdown could be decreased to an average rate of 1,500 kg/h, assuming that it remained at 204°C? How much additional heat would be available from the 1,500 kg/h of blowdown water for use in heating the incoming makeup water?

Given:

Fuel cost:	€65.00	/tonne
Energy content:	28,000	MJ/tonne coal
Initial blowdown rate:	7,000	kg/h
Blowdown temperature:	204	°C
Assume 100% of heat could be used.		

Solution: hw = 871.8 kJ/kg

Assuming blowdowns last for about one hour per day.

Therefore, the annual cost savings (CS) from reducing the blowdown mass flow rate can be estimated as follows:

$$CS1 = hw \times (m1 - m0) \times 365 \ h/yr \times €65/tonne/1,000 \ kg/tonne/28,000 \ MJ/tonne$$
$$= 871.8 \ kJ/kg \times (7,000 \ kg/h - 1,500 \ kg/h) \times 365 \ h/yr \times €65/tonne/1,000 \ kg/tonne/28,000 \ MJ/tonne$$
$$= \textbf{€ 4,062.82/yr}$$

Additionally, the heat (q) available from the 1,500 kg/h of blowdown water can be estimated as follows (assuming 100% recovery of the heat):

$$q = hw \times 1,500 \ kg/h$$
$$= 871.8 \quad kJ/kg \times 1,500 \ kg/h$$
$$= \textbf{1,307,700} \quad \textbf{kg/h of blowdown}$$

Therefore, the annual cost savings (CS2) of recovering all the heat from the blowdowns can be estimated as follows:

$$CS2 = q \times 365 \ h/yr \times €65/ton/1,000 \ kg/ton/28,000 \ MJ/kg$$
$$= 1,307,700 \quad kg/h$$
$$\times 365 \ h/yr \times €65/ton/1,000 \ kg/ton/28,000 \ MJ/kg$$
$$= €1,108/yr$$

Finally, the total annual cost savings (CS) from these two measures is:

$$CS = \textbf{€5,170.86/yr}$$

Problem: Develop a table showing the size of the orifice, the number of kilograms of steam lost per hour, the cost per month, and the cost for an average heating season of 7 months.

Given: Suppose that you are preparing to estimate the cost of steam leaks in a 2.4 MPa steam system. The source of the steam is 28,000 MJ/t coal at €70/tonne, and the efficiency of the boiler plant is 70%. Hole diameters are classified as 2, 4, 6, 8, and 10 mm.

Solution:

Size of leak (mm)	kg/hr lost	Heat loss (kJ/h)	Monthly cost (€)	Annual cost (€)
2	28	75,971	€ 198	€ 1,386
4	112	303,885	€ 792	€ 5,546
6	251	683,742	€ 1,783	€ 12,478
8	447	1,215,541	€ 3,169	€ 22,184
10	699	1,899,282	€ 4,952	€ 34,662

Assume starting temperature of the make-up water is room temperature (20°C)

Problem: If this change is made, how many kilograms per hour of steam does this energy management opportunity (EMO) save?

Given: A 100-metre-long steam pipe carries saturated steam at 670 kPa. The pipe is not well insulated, and has a heat loss of about 50 MJ per hour. The plant industrial engineer suggests that the pipe insulation be increased so that the heat loss would be only 5 MJ per hour.

Solution: The heat rate savings (q) can be estimated as follows:

$$q = q1 - q0$$
$$= 50 \text{ MJ/hr} - 5 \text{ MJ/hr}$$
$$= 45 \text{ MJ/hr}$$

Additionally, the mass flow rate of steam saved (m) can be calculated as follows:

$$m = q/hstm@670 \text{ kPa}$$
$$= 45 \text{ MJ/hr}/2760.7 \text{ kJ/kg}$$
$$= \textit{16.3 kg/h}$$

Problem: Calculate the heat loss from the boiler from the following two sources.

Given: Tastee Orange Juice Company has a large boiler that has a 40 m^2 exposed surface that is at 110°C. The boiler discharges flue gas at 205°C, and has an exposed surface for the stack of 15 m^2.

Solution:

Radiative loss $= A \times 5.6697 \times 10^{-8}$ (W/m^2/K^4) $\times$ (Ts^4 – Tr^4)

and

Convective loss $= A \times 4{,}480$ J/h/m^2/K$^{4/3} \times$ (Ts – Tr)$^{(4/3)}$

where,

Ts $=$ Surface temperature in degrees Kelvin
Tr $=$ Room temperature in degrees Kelvin
A $=$ Surface area in square metres

Therefore, the total heat losses from these two sources can be calculated as follows:

q $=$ 40 m^2 $\times$ [5.6697 $\times$ 10^{-8} J/s/m^2/K $\times$
((273.15 + 110)^4 – (273.15 + 25)^4) $\times$ 3600 s/h
$+$ 4.48 kJ/h/m^2/K$^{4/3}$ $\times$ ((273.15 + 110) – (273.15
$+$ 25))$^{(4/3)}$)K]
$+$ 15 m^2 $\times$ [5.6697 $\times$ 10^{-8} J/s/m^2/K^4 $\times$ ((273.15+205)^4
$-$ (273.15+25)^4) $\times$ 3600 s/h
$+$ 4.48 kJ/h/m^{2}K$^{4/3}$ $\times$ ((273.15+205)
$-$ (273.15+205))$^{(4/3)}$)K]
$=$ *382,548 kJ/h*

Problem: What is the relationship of the wisp, moderate leak, and severe leak as defined by Waterland to the hole sizes found from Grashof's formula for 4.2 kPa steam?

In other words, find the hole sizes that correspond to the wisp, moderate leak, and severe leak.

Given: In Section 8.2.1.1 two methods were given to estimate the energy lost and cost of steam leaks.

Solution:

	Size of leak (mm diameter)	kg/hr lost
Wisp 0.039	28.264	11.3
Moderate	56.527	45.4
Severe	154.807	340.2

Assumed   4.2 kPa saturated steam system

Chapter 11

Control Systems and Computers

Problem: How much will be saved by duty-cycling the fans such that
each is off 10 minutes per hour on a rotating basis?
At any time, two fans are off and 10 are running.

How much will they be willing to spend for a control system
to duty cycle the fans?

Given: Ugly Duckling Manufacturing Company has a series of 12
exhaust fans over its diagnostic laboratories. Presently, the
fans run 24 hours per day, exhausting 280 l/s each. The
fans are run by 1.5-kW motors with load factors of 0.8 and
efficiencies of 80%. Assume the plant operates 24 hours per
day, 365 days per year in an areas of 2,500°C heating degree
days and 1,000°C cooling degree days per year.

The plant pays €0.05 per kWh and €5 per kW for it electricity and €5 per GJ for its gas. The heating plant efficiency is
0.8, and the cooling COP is 2.5. Assume the company only
approves EMO projects with two years or less SPP.

Solution:

$$\text{DR fan} = \text{If} \times 1.5 \text{ kW/fan} \times 2 \text{ fans/eff}$$
$$= 0.8 \times 1.5 \text{ kW/fan} \times 2 \text{ fans/0.8}$$
$$= 3 \text{ kW}$$

$$\text{ES fan} = \text{DR} \times 24 \text{ hrs/day} \times 365 \text{ days/yr}$$
$$= 3 \text{ kW} \times 24 \text{ hrs/day} \times 365 \text{ days/yr}$$
$$= 26{,}280 \text{ kWh/yr}$$

$$\text{CS fan} = \text{ES} \times €0.05/\text{kWh} + \text{DR} \times €5/\text{kW/mo} \times 12 \text{ mo/yr}$$
$$= 26{,}140 \text{ kWh} \times €0.05/\text{kWh} + 3 \text{ kW} \times €5/\text{kW/mo}$$
$$\times 12 \text{ mo/yr}$$
$$= €1{,}494/\text{yr}$$

$$\begin{aligned}
\text{heating savings} &= 2 \text{ fans} \times 280 \text{ l/s/fan} \times 3600 \text{ s/hr} \\
&\quad \times 1.006 \text{ kJ/kg/}°C \times 0.001 \text{ m}^3/1 \times 1.204 \text{ kg/m}^3 \\
&\quad \times 2{,}500°C \text{ days/yr} \times €5/GJ \times 24 \text{ hrs/day} \\
&\quad \times GJ/1{,}000{,}000 \text{ kJ}/0.8 \\
&= €916/\text{yr}
\end{aligned}$$

$$\begin{aligned}
\text{cooling savings} &= 2 \text{ fans} \times 280 \text{ l/s/fan} \times 1.006 \text{ kJ/kg/}°C \\
&\quad \times 0.001 \text{ m}^3/1 \times 1.204 \text{ kg/m}^3 \times 1{,}000°C \text{ days/yr/} \\
&\quad \times 1 \text{ kW/1 kJ/s} \times 24 \text{ h/day} \times \\
&\quad (€0.05/kWh + €5/kW/mo \\
&\quad \times 12 \text{ mo/yr}/8{,}760 \text{ h/yr})/2.5 \\
&= €370/\text{yr}
\end{aligned}$$

Therefore, the total cost savings is:

$$CS = €2{,}780/yr$$

Additionally, they would be willing to pay the following for implementation cost (IC):

$$\begin{aligned}
IC &= SPP \times CS \\
&= 2 \text{ years} \times €2{,}780/\text{yr} \\
&= €5{,}560
\end{aligned}$$

Problem: What is the savings for turning these lamps off an extra 4 hrs/day?

What type of control system would you recommend for turning off the 1,000 lamps? (Manual or automatic? Timers? Other sensors?)

Given: Profits, Inc., has a present policy of leaving all of its office lights on for the cleaning crew at night. The plant closes at 1800 hours, and the cleaning crew works from 1800 to 2200. After a careful analysis, the company finds it can turn off 1,000 40W fluorescent lamps at closing time. The remaining 400 lamps have enough light for the cleaning crew. Assume the company works 5 days/wk, 52 wks/yr, and pays €0.06/kWh and €6/kW for electricity. Peaking hours for demand are 1300 to 1500. Assume there is one ballast for every two lamps and the ballast adds 15% to the load of the lamps.

Solution:

$$CS = 1{,}000 \text{ lamps} \times 40 \text{ W/lamp} \times \text{kW}/1{,}000 \text{ W} \times 4 \text{ h/day} \times 5 \text{ days/wk} \times 52 \text{ wk/yr} \times €0.06/\text{kWh} \times 1.15$$
$$= €2{,}870/yr$$

I would recommend an automatic timer for the 1,000 lamps and possibly occupancy sensors for the other 400 lamps.

Problem: How much did it cost the company in extra charges not to have the lights on some kind of control system?

What type of control system would you recommend and why?

Given: In problem 11.2, assume that the plant manager has checked on the lighting situation and discovered that the cleaning crew does not always remember to turn the remaining lights off when they leave. In the past years, the lights have been left on overnight an average of twice a month. One of the times the lights were left on over a weekend.

Solution: The cost (C) from leaving the 400 lights on 8 hours a night for 23 nights a year and 56 hours on weekend:

$$CS = 400 \text{ lamps} \times 40 \text{ W/lamp} \times \text{kW/1,000 W} \times (8 \text{ h/day} \times$$
$$23 \text{ days/yr} + 56 \text{ h/yr}) \times €0.06/\text{kWh} \times 1.15$$
$$= \boldsymbol{€265/yr}$$

I would recommend an automatic timer for the 1,000 lamps and possibly occupancy sensors for the other 400 lamps.

The timers would turn off the unnecessary lighting when even when the cleaning crew is working, and the occupancy sensors ensure the lights turn off when no one is present.

Problem: What is the savings in kilojoules for this setback?

How could this furnace setback be accomplished?

Given: Therms, Inc., has a large electric heat-treating furnace that takes considerable time to warm up. However, a careful analysis shows the furnace could be turned back from a normal temperature of 1,000°C to 425°C, 20 hours/week and be heated back up in time for production. The ambient temperature is 20°C, and the composite R-value of the walls and roof is 1.7, and the total surface area is 100 m².

Solution: q = U × A × ΔT × time
= 1/R × A × ΔT × time
= (1/1.7) W/(m²°C) × 100 m² × (1,000°C − 425°C) × 20 h/wk × 52 wk/yr × (kJ/(1,000 W s)) × 3,600 s/h
= *126,635,294 kJ/yr*

This furnace setback could be accomplished with a program-mable logic controller (PLC).

Problem: Obtain bin data for your region, and calculate the savings in kJ for a nighttime setback of 8°C from 19°C to 11°C, 8 hours per day (midnight to 0800)

Solution: If we assume the bin data yield 1,000 degree-days, then using Figure 11-1 the savings if the setback occurred 24 hours a day would be 2,556 kJ/m²/yr. Therefore, since 8 h/day is one-third of the time, the saving would be *852 kJ/m²/yr.*

Problem: What is the savings? Determine the SPP.
              Would you recommend it to the company?

Given: Petro Treatments has its security lights on timers. The com-
              pany figures an average operating time of one hour per day
              can be saved by using photocell controls. The company has
              100 mercury vapor lamps of 1,000 Watts each, and the lamp
              ballast increases the electric load by 15%. The company pays
              €0.06/kWh. Assume there is no demand savings. The pho-
              tocell controls cost €10 apiece and each lamp must have its
              own photocell. It will cost the company an average of €15
              per lamp to install the photocells.

Solution: CS = 1000 W/lamp × 100 lamps
                 × kW/1,000 W × 1.15 × 365 h/yr × €0.06/kWh
              = *€2,519/yr*
              SPP = IC/CS
              = (€15+€10)/lamp × 100 lamps/€2,519/yr
              = *0.99 years*

              Since the payback period is less than one year, I would
              recommend this project.

Problem: Would you recommend this change? Why?

Given: CKT Manufacturing Company has an office area with a number of windows. The offices are presently lighted with 100 40-W fluorescent lamps. The lights are on about 3,000 hours each year, and CKT pays €0.08 per kWh for electricity. After measuring the lighting levels throughout the office area for several months, you have determined that 70% of the lighting energy could be saved if the company installed a lighting system with photo sensors and dimmable electronic ballasts and utilized daylighting whenever possible.

The new lighting system using 32-Watt T-8 lamps and electronic ballasts together with the photo sensors would cost about €2,500.

Solution: CS = 40 W/lamp × 100 lamps × kW/1,000 W × 1.15 ×
            3,000 h/yr × €0.08/kWh × 0.7
           = €773/yr
    SPP = IC/CS
           = €2,500/€773/yr
           = 3.23 years

Since the payback period is less than five years, I would recommend this project.

Chapter 12

Maintenance

Problem: Based on this table, give a range of times for possible intervals for changing filters.

Given: In determining how often to change filters, an inclined tube manometer is installed across a filter. Conditions have been observed as follows:

Week	Manometer reading	Filter condition
1	1.0 in water	Clean
2	1.5	Clean
3	1.8	A bit dirty
4	2.0	A bit dirty
5	2.0	A bit dirty
6-9	2.3	Dirty
10-13	2.5	Dirty
14-18	2.8	Dirty
19-23	3.0	Very Dirty
24	3.3	Plugged up: changed

Solution:

One possible interval for changing the filter is once every 14 to 18 weeks.

Problem: Calculate the standard time for filter cleaning.

Given: You have been keeping careful records on the amount of time taken to clean air filters in a large HVAC system. The time taken to clean 35 filter banks was an average of 18 min/filter bank and was calculated over several days with three different people: one fast, one slow, and one average. Additional time that must be taken into account includes personal time of 20 minutes every 4 hours. Setup time was not included. Assume that fatigue and miscellaneous delay have been included in the observed times.

Solution:

ST = 35 filter banks × 18 min/filter bank × (1 + 20 min/240 min)/35 filter banks

= *19.5 min/filter*

Problem: How many people could you have hired for the money you lost?

Given: Your company has suffered from high employee turnover and production losses, both attributed to poor maintenance (the work area was uncomfortable, and machines also broke down). Eight people left last year, six of them probably because of employee comfort. You estimate training costs as €10,000 per person. In addition, you had one 3-week problem that probably would have been a 1-week problem if it had been caught in time. Each week cost approximately €10,000 All these might have been prevented if you had a good maintenance staff. Assume that each maintenance person costs €25,000 plus €15,000 in overhead per year.

Solution: The cost (C) due to poor maintenance conditions can be estimated as follows:

   C = Six people lost due to poor maintenance (comfort) × €10,000/person in training + (3 weeks – 1 week) × €10,000/week of downtime
     = €80,000

Therefore, the number of maintenance people (N) you could hire can be calculated as follows:

   N = C/(salary + overhead)
     = €80,000/(€25,000 + €15,000)/person
     = *2 maintenance people*

Problem: How large an annual gas bill is needed before adding a maintenance person for the boiler alone is justified if this person would cost €40,000 per year?

Given: A recent analysis of your boiler showed that you have 15% excess combustion air. Discussion with the local gas company has revealed that you could use 5% combustion air if your controls were maintained better. This represents a calculated efficiency improvement of 2.3%.

Solution: The annual gas bill required to pay for a maintenance person that would increase boiler efficiency by 2.3% can be calculated as follows:

$$\text{gas bill} = €40,000/\text{yr}/2.3\%$$
$$= €1,739,130.43/yr$$

Problem: What annual amount would this improvement be worth
             considering energy costs only?

Given: Your steam distribution system is old and has many leaks.
             Presently, steam is being generated by a coal-fired boiler,
             and your coal bill for the boiler is €600,000 per year. A
             careful energy audit estimated that you were losing 15% of
             the generated steam through leaks and that this could be
             reduced to 2%

Solution: CS $=$ $(15\% - 2\%) \times$ €600,000/yr
                   $=$ **€78,000/yr**

Problem: Group relamping is a maintenance procedure recommended in Chapter Five. Using data from Chapter Five, construct a graph which plots maintenance cost per hour and relamping interval expressed as a percentage of the lamps rated life against the total relamping cost. Can you construct such a graph that will provide the answer to the question whether group relamping is cost-effective for a particular company?

Solution:

$I = 80\%$ L (/lamp) = €0.85

Maintenance Cost per hour (€) (H)	Relamping Interval of rated life (I)	Product of hourly maintenance cost and (% relamping interval) (H × I)	Relamping Cost per lamp (G)	Maintenance Cost per hour (€) (H)	Total spot relamping cost (€/lamp) (Cs)	Total group relamping cost (€/lamp) (Cg)
10	75%	€7.50	€2.24	10	5.85	2.10
15	80%	€11.25	€2.80	15	8.35	2.63
20	85%	€15.00	€3.36	20	10.85	3.15
25		€18.75	€3.91	25	13.35	3.67
30		€22.50	€4.47	30	15.85	4.19
		€8.00	€2.10			
		€12.00	€2.63			
		€16.00	€3.15			
		€20.00	€3.67			
		€24.00	€4.19			
		€8.50	€1.98			
		€12.75	€2.47			
		€17.00	€2.96			
		€21.25	€3.45			
		€25.50	€3.94			

10.6 Chart 2 constructed with $I=80\%$ and lamp cost of €0.85, in which case, with the given maintenance costs, it is always cost effective to group relamp. Also, assume maintenance time includes that it takes 30 minutes per lamp to spot relamp and 5 minutes per lamp to group relamp.

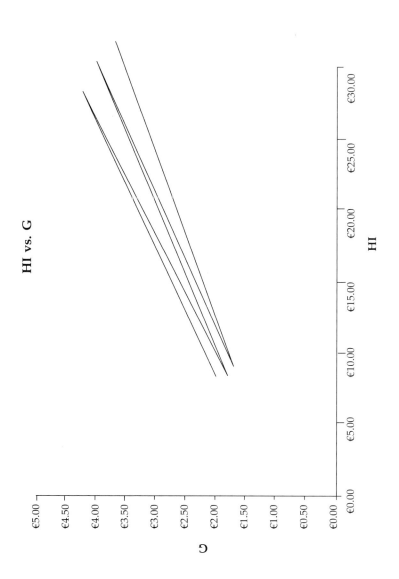

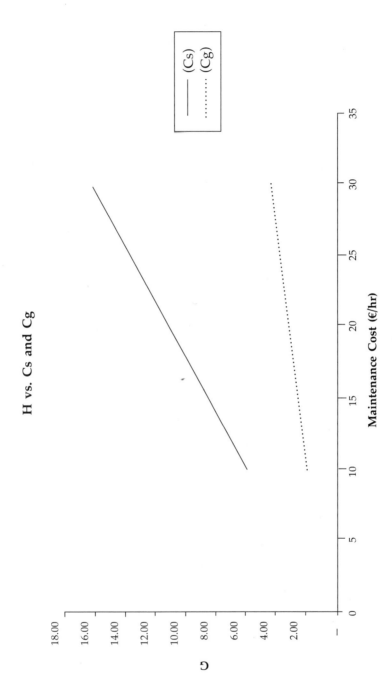

Chapter 13

Insulation

Problem: What is the heat loss per year in kJ? What is the cost of this heat loss?

Given: A metal tank made out of mild steel is 1.3 m in diameter, 2 m long, and holds water at 80C. The tank holds hot water all the time is on a stand so all sides are exposed to ambient conditions at 25°C. The boiler supplying this hot water is 79% efficient and uses natural gas costing €5/GJ. Assume there is no air movement around the tank

Solution: assume thickness of the tank is 1.25 cm

K (W/(m^2 °C))	45.34	
R ((m^2 °C)/W)	0.00028	R=d/K
R surface	0.08	(m^2 °C)/W
U	12.46	W/(m^2 °C)
T(amb)	25	°C
T(inside)	80	°C
A = piDH+2pir^2	11	m^2
Q = UAΔT	7,415	W
Hours per year (h)	8,760	h/yr
q = Qh	*233,842,910*	*kJ/yr*
c	€5.00	/GJ
eff	79%	
C = qc/eff	*€1,480.02/yr*	

Problem: Calculate the present worth of the proposed investment.

Given: Ace Manufacturing has an uninsulated condensate return tank holding pressurized condensate at 140 kPa saturated. The tank is 2.5 m in diameter and 1.3 m long. Management is considering adding 5 cm of aluminum-jacketed fiberglass at an installed cost of €6.50 per m². The steam is generated by a boiler which is 78% efficient and consumes No. 2 fuel oil at €7/GJ. Energy cost will remain constant over the economic life of the insulation of 5 years. Ambient temperature is 20°C. Rs is 0.06 for the uninsulated tank. The tank is used 8,000 h/yr.

Solution:

assume thickness of the tank is	1.25	inch
K (W/(m² °C))	45.34	
R ((m² °C)/W)	0.00028	R=d/K
R surface	0.06	(m² °C)/W
U	16.59	W/(m² °C)
T(amb)	20	°C
T(inside)	109	°C
A = piDH+2pir²	12	m²
Q = UAΔT	18,426	W
Hours per year (h)	8,000	h/yr
q = Qh	530,655,212	kJ/yr
c	€7.00	/GJ
eff	78%	
Co = qc/eff	€4,762.29	/yr

assume thickness of the tank is	1.25	cm
K (W/(m² °C))	45.34	
R ((m² °C)/W)	0.00028	R = d/K
R surface	0.13	(m² °C)/W
K insulation	0.036	W/(m² °C)
d insulation	5	cm
R insulation	1.39	(m² °C)/W
U	0.66	W/(m² °C)
T(amb)	20	°C
T(inside)	109	°C
A = piDH+2pir²	12	m²
Q − UAΔT	730	W
Hours per year (h)	8,000	h/yr
q = Qh	21,028,118	kJ/yr
c	€7.00	/GJ
eff	78%	
Cf qc/eff	€188.71	/yr

Therefore, the annual cost savings (CS) is:

$$CS = Co - Cf = \qquad €4{,}573.58 \text{ /yr}$$

Additionally, the implementation cost (IC) of the insulation installation is:

$$IC = A \times €6.50/m^2 = \qquad €80.83$$

Finally, the present worth (NPV) can be calculated as follows:

$$
\begin{aligned}
P \quad &= A[P \mid A, 15\%, 5] - IC \\
&= €4{,}573.58[3.3522] \; - \; €80.83 \\
&= \boldsymbol{€15{,}250.71}
\end{aligned}
$$

Problem: What is the savings in euros and GJ?

Given: Your plant has 160 m of uninsulated hot water lines carrying water at 80°C. The pipes are 10 cm in nominal diameter. You decide to insulate these with 5-cm calcium silicate snap-on insulation at €10/m² installed cost. The boiler supplying the hot water consumes natural gas at €6/GJ and is 80% efficient. Ambient air is 27°C, and the lines are active 8,760 h/yr.

Solution:

assume thickness	1.25	cm
K (W/(m² °C))	45.34	
R ((m² °C)/W)	0.00028	R = d/K
R surface	0.08	(m² °C)/W
U	12.46	W/(m² °C)
T(amb)	27	°C
T(inside)	80	°C
A = piDH	50	m²
Q = UAΔT	33,187	W
Hours per year (h)	8,760	h/yr
qo = Qh	1,046.57	GJ/yr
c	€6.00	/GJ
eff	80%	
Co = qc/eff	€7,849.27	/yr

assume thickness	1.25	cm
K (W/(m^2 °C))	45.34	
R ((m^2 °C)/W)	0.00028	R = d/K
R surface	0.08	(m^2 °C)/W
K insulation	0.06	W/(m^2 °C)
d insulation	6.82	cm
R insulation	1.18	(m^2 °C)/W
U	0.79	W/(m^2 °C)
T(amb)	27	°C
T(inside)	80	°C
A = piDH	104	m^2
Q = UAΔT	4,376	W/h
Hours per year (h)	8,760	h/yr
qf = Qh	138.01	GJ/yr
c	€6.00	/GJ
eff	80%	
Cf = qc/eff	€1,035.06	/yr

Therefore, the annual cost savings (CS) is:
$$CS = Co - Cf = \quad \textbf{€6,814.21/yr}$$

Additionally, the amount of heat saved (ES) is:
$$ES = qo - qf = \quad \textbf{909 GJ/yr}$$

Problem: What is the cost of heat loss and heat gain per m² for a year?

Given: Given a wall constructed as shown in Figure 11-6. HDD are 2,000 °C-days, while CDD are 1,000 °C-days. Heating is by gas with a unit efficiency of 0.7. Gas costs €6/GJ. Cooling is by electricity at €0.06/kWh (ignore demand costs), and the cooling plant has a 2.5 seasonal COP.

Solution:

Layer	R ((m² °C)/W)	d (cm)	K ((W)/(m² °C))
Outside film	0.30		
Brick	0.08	10	1.30
Mortar	0.02	1.2	0.72
Block	0.13	0	0.79
Plaster board	0.08	1.2	0.15
inside film	0.09		
	0.69		

U (1/R) $\qquad$ 1.45 W/(m²°C)

$$Q/A = U \times DD/yr \times 24 \text{ h/day}$$
$$Q/A \text{ heating} = \frac{1.45 \text{ J}/(\text{s m}^2 \text{ °C}) \times 2{,}000 \text{ °C days/yr} \times 24 \text{ h/day}}{3{,}600 \text{ s/h}}$$
$$= 251{,}247 \text{ kJ/m}^2/\text{yr}$$

$$Q/A \text{ cooling} = \frac{1.45 \text{ J}/(\text{s m}^2 \text{ °C}) \times 2{,}000 \text{ °C days/yr} \times 24 \text{ h/day}}{3{,}600 \text{ s/h}}$$
$$= 125{,}623 \text{ kJ/m}^2/\text{yr}$$

Therefore, the cost of the heating loss (Ch) can be estimated as follows:

$$
\begin{aligned}
Ch &= Q/A \text{ heating} \times €6/GJ/0.7 \\
&= 0.25 \text{ GJ/m}^2/\text{yr} \times €6/GJ/0.7 \\
&= €2.15/\text{m}^2/\text{yr}
\end{aligned}
$$

Additionally, the cost of the cooling loss (Cc) can be estimated as follows:

$$
\begin{aligned}
Cc &= Q/A \text{ cooling} \times kWh/3{,}412 \text{ W} \times €0.06/kWh/2.5 \\
&= 125{,}623 \text{ kJ/m}^2/\text{yr} \times kW/kJ/s \times (1 \text{ h}/3{,}600s) \\
&\quad \times €0.06/kWh/2.5 \\
&= €0.84/\text{m}^2/\text{yr}
\end{aligned}
$$

Finally, the total cost (C) is:

$$
\begin{aligned}
C &= Ch+Cc \\
&= €2.99/\text{m}^2/\text{yr}
\end{aligned}
$$

Problem: How much fiberglass insulation with a Kraft paper jacket is necessary to prevent condensation on the pipes?

Given: A 15-cm pipe carries chilled water at 4°C in an atmosphere with a temperature of 32°C and a dew point of 30°C.

Solution:

$$Q \text{ total} = (32°C\text{-}4°C)/(Rp+Ri+Rs)$$
$$= (32°C\text{-}30°C)/Rs$$
$$= (30°C\text{-}4°C)/Ri$$

$$Rs = 0.09 \ (m^2 \ °C)/W$$

$$Ri = 26°C \times 0.09 \ (m^2 \ °C)/W/2°C$$
$$= 1.21 \ (m^2 \ °C)/W$$

$$di = Ri \times K$$
$$= 1.21 \ (m^2 \ °C)/W \times 0.03605 \ (W)/(m^2 \ °C)$$
$$= \textit{4.37 cm of fiberglass insulation}$$

Problem: What is the R-value of one of the walls with just a window? What is the R-value of the wall with the window and the door? What is the R-value of the roof? How many kJ must that air conditioner remove to keep the inside temperature at 26°C? How many kWh of electric energy will be used in that one-hour period by the air conditioner?

Given: A building consists of fours walls that are each 3 m high and 7 m long. The wall is constructed of 10 cm of corkboard, with 2.5 cm of plaster on the outside and 1 cm of gypsum board on the inside. Three of the walls have 2 × 1.25 m, single-pane windows with R = 0.1. The fourth wall has a 2 × 1.25 m window and a 1 × 2 m door made of 2.5 cm thick softwood. The roof is constructed of 2 cm plywood with asphalt roll roofing over it. The inside temperature of the building is regulated to 26°C by an air-conditioner operating with a thermostat. The air-conditioner has an SEER of 8. The outside temperature is 35°C for one hour.

Solution:

	Area win. (m^2)	Area of door (m^2)	Area of wall (m^2)	Total Area less door, less win. (m^2)
wall with 1 window	2.5		21	18.5
wall with and door	2.5	2	21	16.5
roof				49

	K ((W)/(m² °C))	d (cm)	R = d/K ((m² °C)/W)	Wall w/o win and door	Window	Door	Roof
Corkboard	0.04	10	2.57	2.57			
Plaster	0.72	2.5	0.03	0.03			
Gypsum Board		1	0.06	0.06			
Surface film			0.09	0.09		0.09	0.09
Windows			1.10		1.10		
Softwood	0.12	2.5	1.25			1.22	
Plywood	0.12	2	0.94				0.17
asphalt roll roofing							0.03
				2.76	0.10	0.31	0.29

R-value of the wall with one window:

$$\text{R-value} = 1/[(1/2.76)\,(19/21) + (1/0.1)\,(2/21)]$$
$$(m^2\ °C)/W$$
$$= \textbf{0.66}\ \textbf{\textit{(m}}^2\ \textbf{\textit{°C)/W}}$$

R-value of the wall with one window and one door:

$$\text{R-value} = \textbf{0.56}\ \textbf{\textit{(m}}^2\ \textbf{\textit{°C)/W}}$$

R-value of the roof:

$$\text{R-value} = \textbf{0.29}\ \textbf{\textit{(m}}^2\ \textbf{\textit{°C)/W}}$$

$$
\begin{aligned}
Q\ \text{lost} &= U A \Delta T \\
&= (35°C - 26°C) \times [(1/0.66\ (m^2\ °C)/W) \\
&\quad \times 21\ m^2/\text{wall} \times 3\ \text{walls} \\
&\quad + (1/0.56\ (m^2\ °C)/W) \times 21\ m^2/\text{wall} \times 1\ \text{wall} \\
&\quad + (1/0.29\ (m^2\ °C)/W) \times 49\ m^2] \\
&= 2{,}706\ W \\
&= 2{,}706\ J/s
\end{aligned}
$$

$$q\ lost = \textbf{9,706 kJ} \quad \text{in the one hour}$$

$$
\begin{aligned}
kWh &= 9{,}740\ kJ \quad \times Wh/8.44\ kJ \times kW/1{,}000\ W \\
&= \textbf{1.154 kWh}
\end{aligned}
$$

Problem: Repeat Problem 13.6 with the single-pane windows replaced with double-paned windows having an R-value of 0.16.

Solution:

	Area win. (m²)	Area of door (m²)	Area of wall (m²)	Total Area less door, less win. (m²)
wall with 1 window	2.5		21	18.5
wall with window and door	2.5	2	21	16.5
roof				49

	K ((W)/(m² °C))	d (cm)	R = d/K ((m² °C)/W)	Wall w/o win and door	Window	Door	Roof
Corkboard	0.04	10	2.57	2.57			
Plaster	0.72	2.5	0.03	0.03			
Gypsum Board		1	0.06	0.06			
Surface film			0.09	0.09		0.09	0.09
Windows			1.16		1.16		
Softwood	0.12	2.5	1.22			1.22	
Plywood	0.12	2	0.17				0.17
asphalt roll roofing			0.03				0.03
				2.76	0.16	0.31	0.29

R-value of the wall with one window:

R-value $= 1/[(1/2.76)\ (19/21) + (1/0.16)\ (2/21)]$
$(m^2\ °C)/W$

$= \textbf{0.94 } (m^2\ °C)/W$

R-value of the wall with one window and one door:

R-value $= \textbf{0.75 } (m^2\ °C)/W$

R-value of the roof:

R-value $= \textbf{0.29 } (m^2\ °C)/W$

Q lost $= U A \Delta T$

$= (35°C - 26°C) \times [(1/0.94\ (m^2\ °C)/W)$
$\times 21\ m^2/wall \times 3\ walls$
$+ (1/0.75\ (m^2\ °C)/W) \times 21\ m^2/wall \times 1\ wall$
$+ (1/0.29\ (m^2\ °C)/W) \times 49\ m^2]$

$= \quad 2,368\ W$

$= \quad 2,368\ J/s$

q *lost* $= \textbf{8,525 } kJ$ in the one hour

$kWh = 8,525\ kJ \quad \times Wh/8.44\ kJ \times kW/1,000\ W$
$= \textbf{1.010 } kWh$

Problem: How many euros per year can be saved by insulating the end cap?

What kind of insulation would you select?

If that insulation cost €300 to install, what is the SPP for this EMO?

Given: While performing an energy audit at Ace Manufacturing Company you find that their boiler has an end cap not well insulated. The end cap is 2 m in diameter and 0.7 m long. You measure the temperature of the end cap as 120°C. The temperature in the boiler room averages 32°C, the boiler is used 8,760 h/yr, and fuel for the boiler is €6/GJ. Assume boiler efficiency of 80%.

Solution:

assume thickness of the end cap is   5 cm

assume the end cap is made of mild steel

K (W/(m^2 °C))	45.34	
R ((m^2 °C)/W)	0.00110	R=d/K
R surface	0.07	(m^2 °C)/W
U	13.32	W/(m^2 °C)
T(amb)	32	°C
T(inside)	120	°C
A = piDH+pir^2	8	m^2
Q=UAΔT	8,838	W/h
Hours per year (h)	8,760	h/yr
q=Qh	278.72	GJ/yr
c	€6.00	/GJ
eff	80%	

C = qc/eff   *€2,090.40/yr*

I would use a mineral wool fiber.

SPP = IC/C

      = €300/€2,090/yr

      = **0.14 years**

Problem: What is the most cost effective solution between the two alternatives?

Given: Assume the tank in Problem 13.1 is a hot water tank that is heated with an electrical resistance element. If this were a hot water tank for a residence, it would probably come with an insulation level of R-7. A friend says that the way to save money on hot water heating is to put a timer or switch on the tank, and to turn it off when it is not being used. Another friend says that the best thing to do is to put another layer of insulation on the tank and not turn it off and on. Assume that there are four of you in the residence, and that you use an average of 80 litres of hot water each per day. Assume that you set the water temperature in the tank to 60°C, and that the water coming into the tank is 20°C. You have talked to an electrician, and she says that she will install a timer on your hot water heater for €50, or she will install an R-13 water heater jacket around your present water heater for €25. Assume that the timer can result in saving three-fourths of the energy lost from the water heater when it is not being used. Electric energy costs €0.08 per kWh.

Solution:

R ((m² °C)/W)	7 (m² °C)/W
U	0.14 W/(m² °C)
T(amb)	25 °C
T(inside)	60 °C
A = piDH+2pir²	11 m²
Q = UAΔT	54 W/h
Hours per year (h)	8,760 h/yr
q = Qh	474 kWh/yr
c	€0.08 /kWh
eff	100%
Co = qc/eff	€37.92 /yr
CS timer = 75% × Co	

$$= 75\% \times €37.92/\mathrm{yr}$$
$$= €28.44/\mathrm{yr}$$

SPP timer $= €50/€28.44/\mathrm{yr}$
$\qquad = 1.76 \text{ years}$

R ((m² °C)/W)	7 (m² °C)/W
R blanket	13 (m² °C)/W
U	0.05 W/(m² °C)
T(amb)	25 °C
T(inside)	60 °C
A = piDH+2pir²	11 m²
Q = UA delta T	19 W/h
Hours per year (h)	8,760 h/yr
q = Qh	166 kWh/yr
c	€0.08 /kWh
eff	100%
Cf = qc	€13.27 /yr

Therefore, the annual cost savings (CS) from the jacket is:

$$CS = Co - Cf = €24.65/\mathrm{yr}$$

SPP timer $= €25/€24.65/\mathrm{yr}$
$\qquad = 1.01 \text{ years}$

Therefore, since the jacket costs less and saves about the same amount, install the jacket.

Chapter 14

Process Energy Management

Problem: If Crown Jewels buys a new motor, which one of these incentives should they ask for?

Given: Florida Electric Company offers financial incentives for large customers to replace their old electric motors with new, high efficiency motors. Crown Jewels Corporation, a large customer of FEC, has a 20-year-old 75-kW motor that they think is on its last legs, and they are considering replacing it. Their old motor is 91% efficient, and the new motor would be 95% efficient. FEC offers two different choices for incentives:

*Either €8/kW (for the size motor considered) incentive or;
*A €150/kW (kW saved) incentive.

Solution: Assume a load factor of 0.6

$$P \text{ saved} = 75 \text{ kW} \times 0.6 \times [(1/0.91) - (1/0.95)]$$
$$= 2.08 \text{ kW}$$

€ incentive for kW saved = €312.32
€ incentive for size of motor = €600

Therefore, they should ask for the €8/kW (size of motor) incentive.

Problem: What is the power factor of this motor?

Given: During an energy audit at the Orange and Blue Plastics Company you saw a 75-kW electric motor that had the following information on the

nameplate: 460 v
114 a
3 phase
95% efficient.

Solution:

P (kW) = sq rt (3) × v × i × pf × 0.95

pf = P/(sq rt (3) × v × i)/0.95

= 75 kW/(sq rt (3) × 0.460 kv × 114 a × 0.95

= *0.869*

Problem: Using the data in Table 14-1, determine whether Ruff should purchase the high efficiency model or the standard model motor?

Find the SPP, ROI, and BCR.

Given: Ruff Metal Company has just experienced the failure of a 15-kW motor on a waste-water pump that runs about 3,000 hours a year.

Assume the new motor will last for 15 years and the company's investment rate is 15%.

Solution: Assume energy cost (EC)   €0.05 /kWh
Assume demand cost (DC)   €7.00 /kW/mo
Assume the motor load factor is   0.6
DR = 15 kW × 0.6 × [(1/0.886) – (1/0.923)]
    = 0.41 kW

Therefore, the cost savings (CS) from using the high-efficiency motor over the standard efficiency motor can be calculated as follows:

CS = DR × DC × 12 mo/yr + DR × 3,000 h/yr × EC
    = 0.41 kW × €7/kW/mo × 12 mo/yr + 0.41 kW
      × 3,000 h/yr × €0.05/kWh
    = €95.29/yr

SPP = Cost premium/CS
     = €186/€95.29/yr
     = *1.95 years*

ROI = *51%*

BCR = PV benefits/PV cost
     = €557.17/€186
     = *3.00*

Therefore, buying the high efficiency motor seems to be a good investment.

Problem: How would you estimate the amount of waste heat that could be recovered for use in heating wash water for metal parts?

Given: A rule of thumb for an air compressor is that only 10% of the energy the air compressor uses is transferred into the compressed air. The remaining 90% becomes waste heat. You have seen a 35-kW air compressor on an audit of a facility, but you do not have any measurements of air flow rates or temperatures.

Solution: Assume that the motor efficiency is 91.5%.
Assume that the compressor motor load factor is 0.6.
Additionally, assume that 80% of the waste heat can be re-covered. Therefore, one can calculate the amount of waste heat available as follows:

$$\begin{aligned}
Q &= \text{lf} \times P \times 90\% \times 80\%/91.5\% \\
&= 0.6 \times 35 \text{ kW} \times 90\% \times 80\%/91.5\% \\
&= 17 \text{ kW} \\
&= \textit{17 kJ/s} \\
&= \textit{520,525 GJ/yr}
\end{aligned}$$

Problem: How much would this load shifting save Orange and Blue
              Plastics on their annual electric costs?

Given: Orange and Blue Plastics has a 110-kW fire pump that must
              be tested each month to insure its availability for emergency
              use. The motor is 93% efficient, and must be run 30 min-
              utes to check its operations. The facility pays €7/kW for
              its demand charge and €0.05/kWh for energy. During your
              energy audit visit to Orange and Blue, you were told that
              they check out the fire pump during the day (which is their
              peak time), once a month. You suggest that they pay one of
              the maintenance persons an extra €50 a month to come in
              one evening a month to start up the fire pump and run it
              for 30 minutes.

Solution:
              Assume the motor load factor is 0.6

              $$DR = 110 \text{ kW} \times 0.6 \times 1/0.93$$
              $$= 70.97 \text{ kW}$$

              Therefore, the electric cost savings (CS) from load shifting
              can be calculated as follows:

              $$CS = DR \times DC \times 12 \text{ mo/yr}$$
              $$= 70.97 \text{ kW} \times €7/\text{kW/mo} \times 12 \text{ mo/yr}$$
              $$= €5,961.29/yr$$

              It would cost €600/yr in extra labor cost. Therefore, the
              total annual savings of over €5,300 makes this look like a
              good EMO.

Problem: What is the implied efficiency of a motor if we say its load
             is 1 kW per kW?

             What is the implied COP of an air conditioner that has a
             load of 1 kW per tonne?

Given: Our "rules of thumb" for the load of a motor and air con-
             ditioner have implicit assumptions on their efficiencies.

Solution:

$$eff = 1 \text{ kW} / 1 \text{kW}$$
$$= \textbf{100\%}$$
$$COP = 1 \text{ tonne/kW} \times 12{,}660 \text{ kJ/tonne-h} \times \text{kW/kJ/s}$$
$$\times 1 \text{ h}/3{,}600 \text{ s}$$
$$= \textbf{3.52}$$

Problem: Even though the motor is expected to last another five years, you think that the company might be better off replacing the motor with a new high-efficiency model. Provide an analysis to show whether this is a cost-effective suggestion.

Given: During an audit trip to a wood products company, you note that they have a 35-kW motor driving the dust collection system. You are told that the motor is not a high efficiency model, and that it is only 10 years old. The dust collection system operates 6,000 hours each year.

Solution:

Since cost premiums range from 10% to 30%, we use 20% to calculate the cost of the high efficiency motor from the premium column in Table 14-1. Therefore, the cost of a 35-kW high-efficiency motor is 5 times €469:

€2,345

Assume energy cost (EC)   €0.05   /kWh
Assume demand cost (DC)   €7.00   /kW/mo
Assume the motor load factor is   0.6
DR $= 35 \text{ kW} \times 0.6 \times [(1/0.915) - (1/0.938)]$
$= 0.56 \text{ kW}$

Therefore, the cost savings (CS) from using the high-efficiency motor over the standard efficiency motor can be calculated as follows:

CS $= \text{DR} \times \text{DC} \times 12 \text{ mo/yr} + \text{DR} \times 6,000 \text{ h/yr} \times \text{EC}$
$= 0.6 \text{ kW} \times €7/\text{kW/mo} \times 12 \text{ mo/yr} + 0.6 \text{ kW}$
$\times 6,000 \text{ h/yr} \times €0.05/\text{kWh}$
$= €216.10/\text{yr}$

SPP $= \text{Cost/CS}$
$= €2,345/€216.10/\text{yr}$
$= \textbf{\textit{10.9 years}}$

ROI $= \textbf{\textit{5.28\%}}$
NPV $= \textbf{\textit{(€998.35)}}$ assuming a MARR of 15%

Therefore, it seems to be a bad project to change the motor now.

Chapter 15

Renewable Energy Sources and Water Management

Problem: How many litres of water would be required to store 1 GJ?

Given: In designing a solar thermal system for space heating, it is determined that water will be used as a storage medium. Assuming the water temperature can vary from 25°C up to 60°C.

Solution:

$$Q = Mc\Delta T)$$
$$M = Q/c\Delta T)$$
$$= 1 \text{ GJ}/(4.184 \text{ kJ/kg °C}) \times (60 \text{ °C} - 25 \text{ °C}))$$
$$= 6{,}829 \text{ kg} \quad (998 \text{ kg/m}^3)$$
$$= \textbf{6,842 L} \quad (1/0.001 \text{ m}^3)$$

Problem: Design the necessary array but neglect any voltage-regulating or storage device.

Given: In designing a system for photovoltaics, cells producing 0.5 volts and 1 ampere are to be used. The need is for a small dc water pump. Drawing 12 volts and 3 amperes.

Solution:

Three branches with 24 cells in each branch:

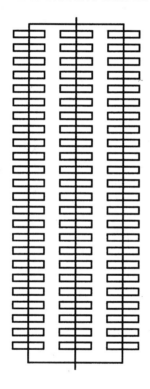

Problem: Calculate the annual water savings (litres and euros) and
annual energy savings (GJ and euros) if the water could be
used as boiler makeup water.

Given: A once-through water cooling system exists for a 75-kW air
compressor. The flow rate is twelve l/min. Water enters the
compressor at 18°C and leaves at 105°C. Water and sewage
cost €0.40/1,000 litres and energy costs €5/GJ. Assume the
water cools to 32°C before it can be used and flows 8,760
h/yr.

Solution: Assume the efficiency of the heating system is 70%.

$$V = 12 \text{ l/min} \times 60 \text{ min/h} \times 8{,}760 \text{ h/yr}$$
$$= 6{,}307{,}200 \text{ l/yr} \qquad (998 \text{ kg/m}^3)$$
$$m = 6{,}294{,}586 \text{ kg/yr} \qquad (1/0.001 \text{ m}^3)$$

$$CS = 6{,}307{,}200 \text{ l/yr} \times €0.40/1{,}000 \text{ l}$$
$$+ 6{,}294{,}586 \text{ kg/yr} \times 4.184 \text{ kJ/kg/°C}$$
$$\times (32°C - 18°C) \times €5/GJ/0.7$$
$$= \textbf{€5{,}156.53/yr}$$

Problem: What is the net annual savings if the sawdust is burned?

Given: A large furniture plant develops 10 tonnes of sawdust (6,000 kJ/tonne) per day that is presently hauled to the landfill for disposal at a cost of €10/tonne. The sawdust could be burned in a boiler to develop steam for plant use. The steam is presently supplied by a natural gas boiler operating at 78% efficiency. Natural gas costs €5/GJ. Sawdust handling and in-process storage costs for the proposed system would be €3/tonne. Maintenance of the equipment will cost an estimated €10,000 per year. The plant operates 250 days/yr.

Solution: Assume the efficiency of the sawdust burning boiler is 70% of the efficiency of the natural gas burning boiler.

$$m = 10 \text{ tonnes/day} \times 250 \text{ day/yr}$$
$$= 2,500 \text{ tonnes/yr}$$

$$CS = 2,500 \text{ tonnes/yr} \times (0.7 \times 6,000 \text{ kJ/tonne}$$
$$\times \text{€5/GJ}$$
$$+ (\text{€10/tonne} - \text{€3/tonne})) - \text{€10,000/yr}$$
$$= \textbf{€7,552.50/yr}$$

Problem: At 40 degree N latitude, how many square feet of solar collectors would be required to produce each month of the energy content of
a) one barrel (160 l) of crude oil?
b) one tonne of coal?
c) 10 m^3 of natural gas?

Solution: Using Table 15-1, look up the data for 40 degree N latitude averages:

6.8 GJ/m^2/yr

Assume that the efficiency of the cell is 0.7 the efficiency of the present fuel

a) A = 5,100,000 kJ/barrel of crude oil/mo
     × 12 mo/yr/6.8 GJ/m^2/yr 0.7
   = *128 m^2*

b) A = 29,079,383 kJ/tonne of coal/mo
     × 12 mo/yr/6.8 GJ/m^2/yr/0.7
   = *73 m^2*

c) A = 372,528 kJ/m^3 of nat. gas/mo
     × 12 mo/yr/6.8 GJ/m^2/yr/0.7
   = *0.9 m^2*

Problem: Determine whether Munich, Dublin, or Bern has the greatest amount of solar energy per square metre of collector surface?

Given: Use Table 13.1. Assume each collector is mounted at the optimum tilt angle for that location.

Solution:

Average Daily Radiation (kJ/day/m²)

City Dublin	Slope	Jan	Feb	Mar	Apr	May	Jun	Jul	Aug	Sep	Oct	Nov	Dec	Total (kJ/day/m²)
	hor	6564	9903	15002	16978	21452	22622	23451	20146	16012	11413	65639	5769	
	30	11527	14854	19124	18193	20850	21043	22247	20782	18965	16205	10686	10686	
	40	12651	15819	19612	17818	19828	19749	20986	20112	19079	17057	11583	11833	
	50	13446	16376	19612	17057	18420	18102	19328	18999	18749	17477	12185	12674	
	vert	13048	14525	15058	10823	10096	9358	10107	11231	13310	14865	11470	12594	
Average Monthly Radiation		1E+07	1E+07	1E+07	1E+07	2E+07	2E+07	2E+07	2E+07	2E+07	1E+07	5E+07	9E+06	*197,285,045*

Average Daily Radiation (kJ/day/m²)

City Bern	Slope	Jan	Feb	Mar	Apr	May	Jun	Jul	Aug	Sep	Oct	Nov	Dec	Total (kJ/day/m²)
	hor	8949	10834	14025	17239	18795	18545	17455	17409	16024	14945	11629	8279	
	30	12049	13196	15399	16978	17023	16217	15547	16535	16921	18216	15922	11459	
	40	12560	13423	15206	16171	15774	14865	14343	15569	16478	18465	16626	12015	
	50	12776	13332	14672	15036	14264	13287	12912	14298	15683	18284	16921	12288	
	vert	10720	10209	9619	8165	6802	6201	6223	7348	9573	13503	14082	10550	
Average Monthly Radiation		1E+07	9E+06	1E+07	1E+07	1E+07	1E+07	1E+07	1E+07	1E+07	1E+07	1E+07	9E+06	*142,914,911*

Average Daily Radiation (kJ/day/m²)

City Munich	Slope	Jan	Feb	Mar	Apr	May	Jun	Jul	Aug	Sep	Oct	Nov	Dec	Total (kJ/day/m²)
	hor	5803	8279	12242	15217	19737	20862	20737	17773	14252	9948	6053	4974	
	30	9426	11595	14911	16058	19045	19317	19556	18091	16455	13446	9289	8358	
	40	10221	12197	15138	15660	18079	18113	18431	17443	16467	14014	9971	9119	
	50	10754	12503	15013	14945	16773	16592	16966	16444	16092	14241	10402	9653	
	vert	10164	10788	11311	9437	9199	8619	8971	9732	11277	11856	9562	9312	
Average Monthly Radiation		8E+06	8E+06	1E+07	1E+07	1E+07	2E+07	2E+07	1E+07	1E+07	1E+07	7E+06	7E+06	*134,942,518*

Therefore, Dublin has greatest amount of solar energy per square metre of collector surface.

Problem: How many litres of gasoline is this?
Using the maximum contents shown in Table 13-15, how many kilograms of corn cobs would it take to equal the gasoline needed to run the car for one year? Rice hulls? Dirty solvent?

Given: A family car typically consumes about 70 GJ per year in fuel.

Solution:

$$V = 70 \text{ GJ/yr} \times 1 \text{ l gasoline}/34{,}838 \text{ kJ}$$
$$= \textit{2009.3 l/yr}$$

Rice hulls:
$$m = 70 \text{ GJ/yr} \times \text{kg}/15{,}118 \text{ kJ}$$
$$= \textit{4{,}630 kg/yr}$$

Corn cobs:
$$m = 70 \text{ GJ/yr} \times \text{kg}/19{,}304 \text{ kJ}$$
$$= \textit{3{,}626 kg/yr}$$

Dirty solvent:
$$m = 70 \text{ GJ/yr} \times \text{kg}/37213 \text{ kJ}$$
$$= \textit{4{,}375 kg/yr}$$

Problem: Determine the power outputs in Watts per square metre for a good wind site and an outstanding wind site as defined in Section 13.5.

Solution: Assume 50% efficiency
             Assume dry air

$$P/A = 0.5 \times eff \times \text{density of air} \times velocity^3$$

Good site: V = 5.8 m/s

$$P/A = 0.5 \times 50\% \times 1.204 \text{ kg/m}^3 \times (5.8 \text{ m/s})^3$$
$$= 59.06 \text{ W/m}^2$$

Outstanding site: V = 8.5 m/s

$$P/A = 0.5 \times 50\% \times 1.204 \text{ kg/m}^3 \times (8.5 \text{ m/s})^3$$
$$184.40 \text{ W/m}^2$$

Problem: How much difference—in percent—is there between the
                 two sites in Problem 13-9?

Solution:

$$P/A = 0.5 \times \text{density of air} \times \text{velocity}^3$$

Good site: V $= 5.8 \text{ m/s}$

$$P/A = 0.5 \times 50\% \times 1.204 \text{ kg/m}^3 \times (5.8 \text{ m/s})^3$$
$$= 59.06 \text{ W/m}^2$$

Outstanding site: V $= 8.493 \text{ m/s}$

$$P/A = 0.5 \times 50\% \times 1.204 \text{ kg/m}^3 \times (8.5 \text{ m/s})^3$$
$$= 184.40 \text{ W/m}^2$$

% difference $= (\text{outstanding} - \text{good})/\text{outstanding}$
                $= 68\%$

or

% difference $= (\text{outstanding} - \text{good})/\text{good}$
                $= 212\%$

Supplemental

Problem: What is the load factor?

Given: A three-phase 50-kW motor draws 27 amps at 480 volts.

   It is 92% efficient and has a reactive power of 10 kVAR.

Solution:

Apparent power $= (3^{0.5}) \times v \times i$
                 $= (3^{0.5}) \times 480 \text{ V} \times 27 \text{ amps}$
                 $= 22.45 \text{ kVA}$

Reactive power $= 10 \text{ kVAR}$

sin (theta) $= 10 \text{ kVAR}/22.45 \text{ kVA}$
theta $= 0.462$

pf $= \cos \text{(theta)}$
   $= 0.895$

Real power $= (3^{0.5}) \times v \times i \times \text{pf}$
   $= 20.10 \text{ kW}$ which is the power actually used

Rated power $= 50 \text{ kW} \times 0.92$
   $= 54.35 \text{ kW}$

Load factor $= 20.1 \text{ kW}/54.35 \text{ kW}$
   $= \mathbf{37.0\%}$

Problem: Compute the monthly facility electric load factor (FLF)

Given: Peak kW   1,250   kW
          Energy use   500,000   kWh
          Time   720   hours

Solution:

   FLF = Actual kWh used/(peak kW × time)
        = 500,000 kWh/(1,250 kW × 720 hours)
        = 55.56%

Problem: How much does it cost to cool 1 million m³ of air from 35°C and 70% relative humidity to 13°C and 95% relative humidity? (AC COP is 2.7 and electricity costs €0.10/kWh)

Solution:

$$\Delta h = (119 - 54) \text{ kJ/kg}$$
$$= 65 \text{ kJ/kg}$$

$$Cost = (10^6 \text{ m}^3) \times (1.204 \text{ kgs/m}^3) \times (65 \text{ kJ/kg})$$
$$\times (\text{kW/kJ/s} \times 1\text{h}/3600\text{s}/2.7 \times (€0.1/\text{kWh})$$

$$= €805.14$$